LEEDS COLLEGE OF BUILDING
WITHDRAWN FROM STOCK

D1326297

ELEMENTS OF ADMINISTRATION
FOR BUILDING STUDENTS

fourth edition

LDS COLL

LEEDS COLLEGE OF BUILDING LIBRARY

CLASS No. 690.068 BUT

ACC. No. 872262

T03952

ELEMENTS OF ADMINISTRATION FOR BUILDING STUDENTS

FOURTH EDITION

John T. Butler

BSc., MCIOB, MIBM, FCSI, MMGB, Dip. Ed.

Head of Department of Construction and Land Administration, Brunel Technical College, Bristol

Hutchinson

London Melbourne Auckland Johannesburg

Hutchinson & Co. (Publishers) Ltd
An imprint of the Hutchinson Publishing Group
17-21 Conway Street, London W1P 6JD

Hutchinson Group (Australia) Pty Ltd
30-32 Cremorne Street, Richmond South, Victoria 3121
PO Box 151, Broadway, New South Wales 2007

Hutchinson Group (NZ) Ltd
32-34 View Road, PO Box 40-086, Glenfield, Auckland 10

Hutchinson Group (SA) (Pty) Ltd
PO Box 337, Bergvlei 2012, South Africa

First published 1970
Reprinted 1974
Second edition 1977
Reprinted with amendments 1979
Third edition 1982
Reprinted 1983
Fourth edition 1988

© John T. Butler 1970, 1977, 1979, 1982, 1988

Printed in Great Britain by The Anchor Press Ltd
and bound by Wm Brendon & Son Ltd
both of Tiptree, Essex

British Library Cataloguing in Publication Data
Butler, John T.
    Elements of administration for building students.
    – 3rd ed.
    1. Construction industry – Great Britain
    – Management
    I. Title
    690′ .068   HD9715.G72

ISBN 0 09 149451 6

Contents

Acknowledgements

The author is grateful to the following for the use of copyright: The Royal Institute of British Architects, The National Joint Council for the Building Industry, The Federated Employers Press Limited.

Special thanks must be offered to the well-known building firm of Morris and Jacombs Limited, Birmingham, for the use of documents, to their members of staff for considerable assistance, and Miss Joan Pinkney for help in typing and Mr Adrian Stadden for his chapter on computers.

Finally the author wishes to offer his most sincere gratitude to his wife for all the help and encouragement she gave to him over the period of preparation.

Preface

It must always be remembered that building is a team challenge and to meet this challenge a clear communication of information is a vital factor of the spoken or written word, drawings or visual charts; but it is essential that all communications made are backed up with a thorough knowledge of the basic principles in addition to an understanding of the end results.

The skills involved are those that require an understanding of organisation and management techniques that may be used for the successful management of a small firm in the construction industry. The work of a technician involves production planning, method study, quality control, etc. This book will provide the basic foundation of organisation in administering principles which the building technician will find valuable both in his course of study and in his day-to-day work, thus giving him a detailed specialist knowledge related to the building processes so creating efficiency in the communication and administration of information in building procedures. It will also be invaluable to students of construction subjects, whatever their particular specialism, who need to have an overall view of the industry and its administration processes, and it will be particularly suitable for students preparing for the Technician Certificate in Building and Civil Engineering Studies and as a foundation study for professional institutes' examinations.

C. H. TUCKER, M.SC., M.C.I.O.B., F.I.A.S., F.C.S.I., M.B.I.M.
Head of Department of Building and Surveying
Reading College of Technology

1   The construction industry

The construction industry is one of the biggest employers of labour in the country, with a labour force of just under one million and an output of £28 billion per year, 80% of this being directly related to the building industry.

The industry as a whole is made up of many fragmented units, about 160 660 in total. The majority of these firms employ less than 25 people, nearly 155 500 of them – with about 64 500 of these being self employed. Middle size companies form less than 5% of the total number indicated from a national survey; those with 25–114 employees totalled 4246 firms; those with 115–599 employees, 717 firms. Large firms employing 600 employees or more accounted for only 111 of the total firms nation wide, only 39 of these employing a work force of over 1200. However, although the number of these large companies represents less than 1% of the industry's organisations, they do constitute a large percentage of its work force and national turnover.

The backbone of the industry is therefore made up of a large number of small concerns, whose main role in the industry is that of repair and maintenance work and new small building projects; functions that are vital for the upkeep of homes, places of work and pleasure. The larger organisations contend in the highly competitive industry for the many and varied types of projects – housing, factories, flats, new town developments and the more complex 'one off' buildings. Besides this type of work many of the larger firms in particular are carrying out major projects overseas, in Europe and the Middle East, in ever-increasing volume.

Progress

The building industry has, over the past few years, entered into a new and challenging era. Many new materials have been developed and are appearing on the market in an ever-increasing volume. New techniques in building in the shape of industrialisation and prefabrication have evolved, bringing with them new problems in design, organisation, production, handling and storage. Plant for use by the builder has kept abreast of the industry, allowing an increase in production without an increase in labour. It should also be realised that the industry is changing within itself. The days when a building contractor would provide his own men to complete a project throughout are now past. Few general building contractors now operate in this way, only perhaps the small builder in the country districts: the reason being, as in most industries, the growth of specialisation or, as more generally known in the industry, sub-contracts.

Sub-contractors

These are smallish groups that specialise in one particular field of construction, for example:

Plastering contractors   Scaffolding contractors
Flooring specialists   Glazing contractors
Painting contractors   Roofing contractors
Demolition contractors   Reinforced concrete
Plant hire firms   specialists

This system of sub-contracting work out generally operates very well, for the firms become experts in their own field, resulting in a high standard of work at very competitive prices. The main difficulty arising out of this is one of co-ordinating and controlling to achieve planning and time schedules.

Labour only (714)

This is at present an expanding group of workers, many becoming self employed and obtaining a Subcontractors Tax Certificate (714). To obtain a certificate so that a contractor will be able to make payments without taking any tax, the self employed must be working in the construction industry in the UK continuously for five years in the six years before applying for a certificate, as well as having a satisfactory record of employment, having a good record of paying Tax and NI – besides running a business properly.

Labour-only Sub-contractors now pay a 2% levy to the CITB, but at present do little training themselves although Government, Unions, industry and the CITB are looking into this problem.

Tradesmen

The construction industry is one of the few that still relies a great deal on the individual skill of a workman. There is, and always will be, the need for operatives such as:

Bricklayers	Slaters and tilers
Carpenters/Joiners	Plumbers
Plasterers	Electricians
Floor and wall tilers	Woodmachinists
Glaziers	Painters and Decorators

and many others who are generally trained through technical colleges or training schemes, to take craft examinations at the end of the apprenticeship period. Many of the better craftsmen with leadership ability go on to posts of importance, for example, general foreman, clerk of works, site agents, etc.

Technicians

There has always been a group of workers in the industry which operates in the area between the professional and management staff, going under the title of either assistant, junior, trainee or some other designation. All these have now come under the heading of technician, and not only do they fulfil a vital role within the industry but the jobs they carry out are wide and varied in content.

Management

It is not only in the field of the tradesman that new methods have developed but also in the overall concept of management, from the board of directors to the foreman on the site. Management staff include:

Construction managers	Site engineers
Contract managers	Surveyors
Estimators	Foremen
Planners	Office managers
Buyers	Accountants
Plant managers	Personnel managers

Training for these positions, and management in general, is now widespread and most builders, including the efficient 'small men', are taking full advantage of it. The chartered Institute of Building is amongst the forerunners in this, although many in the industry are obtaining higher qualifications in the form of degrees.

Professional

This term is used to signify a person working within a chosen profession and includes:

Architects	Civil engineers
Quantity surveyors	Structural engineers
Service engineers	

These persons generally are university graduates or are similarly qualified and carry out most of the design, planning and general stability and serviceability of structures. Professional qualifications depend on their area of work but include membership of R.I.B.A., R.I.C.S. and I.C.E.

2 Parties involved in the building process

Many people are involved in the complex operation of building but these can be classified broadly into three groups:
1 The client, who may be an individual, a local authority, a club committee, a government department or a hospital board.
2 The architect, or in some cases an engineer, whose role is to interpret the client's requirements into a specific design or scheme and generally take over the task of seeing that they are carried through to their conclusion.
3 The general building contractor who may call upon the services of sub-contractors and specialists, who turn the architect's dreams into reality.

This is a very broad outline of the parties concerned in a building project and to understand more fully the workings of the industry, a look in more detail is required. Not only is it necessary to know who these people are (see Fig. 1) but a clear understanding of their duties and responsibilities is essential.

Duties and responsibilities of the client

Commencement
  a His first task is to analyse and collect all the relevant

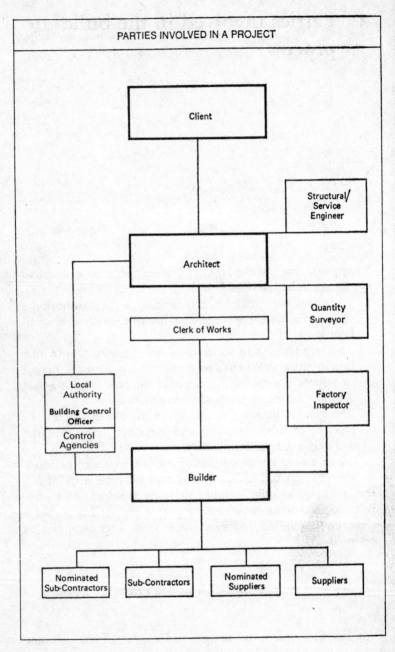

Fig. 1

information that is available to him in order to give the designer the clearest and broadest picture of his requirements with special thought to: I space available; II location; III use of new building; IV cost; V time limits.

b He must consider his legal responsibilities regarding the land, freehold or leases, most likely calling upon the services of a solicitor at this point.

c The financial aspects of the proposed project have to be carefully considered, with consideration being made to drawing regular sums from his account to meet the contractor's payments.

d Engage a designer, usually the architect, to carry out his wishes.

e Ensure that architect is in possession of all the known facts.

f Upon receiving architect's views on type, size, cost of building, consider whether project should proceed.

g If project proceeds, sign contract after agreeing to tender figure.

h Any necessary insurances required by client must be arranged.

During contract

i At the relevant times as stated in the form of contract, meet necessary contractual payments by honouring certificates presented by contractor.

Completion

j Honour the necessary payments of the penultimate certificates, extra contractual claims (if satisfied that they are accurate) and professional fees.

k Make settlement on final certificate.

Duties and responsibilities of the architect

It must first be remembered that on many occasions the client will have little or no knowledge of the industry and in many cases little idea as to what he wants, or even if he has ideas they may be quite inadequate to meet his needs. This must be the architect's first task.

Commencement

a Formulating the client's requirements into an understandable form, bearing in mind any statutory conditions that may apply, such as the Town and Country Planning Acts. It will be advantageous to the client at this stage if he can be shown work of a similar nature so that he may obtain a visual impression of shape, type of materials, size and so on. If this is not possible, pictorial sketches and/or models can be used, but it is often difficult for the client to visualise the true structure from these very artistic creations.

b Bring together a team to give guidance on specific points such as a structural engineer, quantity surveyor, internal services.

c Upon assessing cost limits and time scale, sketch plans can be prepared, enabling client to approve or otherwise, before more final drawings are prepared. The cost of the building will have been broken down into elements at this stage with approximate values so that if costs are to be cut, it can be done in sections, e.g. structure, internal finishings. At the same time, more careful thought is given to any statutory requirements.

d When general agreement has been reached between the architect, his team and the client, the architect can now proceed to produce his contract drawing to enable tenders to be obtained.

e Project drawings complete and bills of quantities prepared, the work can now be put out to tender.

f Upon receipt of checked tenders and the acceptance of one, an inaugural meeting is arranged to clear up salient points that may arise, including a pre-planning period before work on the building is started.

Contract period

g Keep client informed as to project progress and issue architect instructions as necessary.

h Pass on interim certificates of payments to builder (period as stated in formal contract).

Completion

i Upon completion of project, issue certificate of completion.

j Issue certificate of making good defects, normally six months after completion.

Duties and responsibilities of the clerk of works

(The architect's representative on site)

The JCT standard form of building contract states:
'*Clause 12*

'The employer shall be entitled to appoint a clerk of works whose duty shall be to act solely as inspector on behalf of the employer under the directions of the architect, and the contractor shall afford every reasonable facility for the performance of that duty. If any directions are given to the contractor or to his foreman upon the works by the clerk of works the same shall be of no effect unless given in regard to a matter in respect of which the architect is expressly empowered by these conditions to issue instructions and unless confirmed in writing by the architect within two working days of their being given. If any such directions are so given and confirmed, then as from the date of confirmation, they shall be deemed to be architect's instructions'. It is therefore clear that the clerk of works' function is to inspect quality of work in accordance with drawing and specification. He will make regular reports to the architect and it is important that he keep a diary throughout the project which will be invaluable in the case of any disputes arising. He should check setting out and also check other parts of the work during construction to ensure they conform to requirements.

The clerk of works has, on paper, no authority, but as he will have the backing of the architect on most occasions, the builder will usually bend to his wishes and should treat him as a senior member of the project team seeking help and advice as to interpretation of architect's design.

Duties and responsibilities of the quantity surveyor

It must be remembered that although the quantity surveyor is a member of the architect's team from very early on, he

9

must have a very close relationship with the builder upon acceptance of the tender, for it will be his place to cost valuations, variations and the like and, in so doing, remain completely impartial and without favour to either side and so produce harmony in his role of project accountant.

Commencement

a The quantity surveyor is called upon in the early stages of consultation with the client because of his knowledge of costs.

b Prepare an approximate cost from sketch drawings, assembling element costs, so that, should the client's sum be exceeded, the architect can consider each element of the building in reasonable isolation, enabling him to pare costs as necessary.

c Upon acceptance by the client of costs and scheme, the quantity surveyor's next task upon receipt of architect's drawing is to prepare a bill of quantities in accordance with the current Standard Method of Measurement. This is a very important section of the quantity surveyor's responsibilities and great care is taken to ensure accuracy.

d The contractors selected for tendering will each receive a copy of unpriced bills, together with drawings upon which to estimate the project costs. Upon receipt of the tender and now priced bill, the quantity surveyor must check the accuracy to ensure that the builder has made no serious errors which could cause complications at a later date.

e Architect informed of estimate correctness.

Contract period

f The quantity surveyor will carry out monthly valuations, pricing of variation orders and so on in conjuction with the contractor who is allowed to be present at such times, enabling him to receive payment from the client via the architect's interim certificates at regular intervals.

It is at this stage that the contractor must have confidence in the independence of the quantity surveyor in defining work complete (of satisfactory nature) and materials on site (those necessary for work to continue).

g Must keep architect informed as to running cost of project.

Completion

h Prepare, with the aid of builders' receipts and other documents, the final accounts.

i Assist the architect in discussions with the builder as to extra contractual costs.

Duties and responsibilities of structural and service engineer

a Part of the architect's team whose responsibility is to help in the design of the project within the scope of their specialist fields.

b Produce calculations or other relevant data that may be required to assist the architect in his design, the quantity surveyor in his cost control, or the local authority in its assessment of the suitability of the project regarding statutory requirements.

c During the contract, be prepared to assist, modify or re-design work as may become necessary.

Duties and responsibilities of the builder

Commencement

a The builder's first responsibility upon receiving the application to tender is to decide whether or not, with the resources, manpower and general management and administrative set up at his command, he is in a genuine position to take up the project if successful, complete it according to the architect's wishes as stated in the bill/specification and drawing.

b It is to the builder's advantage to carry out, before the tender is complete, a full site investigation so that any unforeseen problems can be noted to ensure that they will not cause delays or extra costs once the contract is under way.

c On completion of the estimate, which will have been obtained from accurate study of the bill of quantities, drawing, site investigation, suppliers' and sub-contractors' quotations, the board of directors or principal of the firm will decide upon the final tender figure. This will then be sent to the architect for his decision on acceptance or refusal according to other tender prices submitted.

d If tender is accepted, the architect's quantity surveyor will require a copy of the priced bills to enable him to satisfy the architect that no serious omissions or faults have been made in the make-up of the tender figure. If a fault or error is found, the builder will be given the opportunity to correct it or, if satisfied with the price, he may allow it to stand.

e Upon the architect and the builder being satisfied with the price, a meeting is arranged between all parties concerned to discuss arrangements, conditions of contract and for the signing of same, when all amendments, disagreements and minor problems have been solved.

During contract

f The builder will be allowed a period for pre-construction planning enabling him to call upon his resources and plan his work and that of any sub-contractors for the contract duration.

g Upon completion of the master programme a meeting should be arranged with the architect, sub-contractors and other interested parties to discuss the programme and agree it. By doing this everyone is put in the picture and their importance shown for the smoothly controlled and efficient running of the job.

h The main responsibility of the builder now is to control labour, including sub-contractors, and materials for the erection of the project, ensuring that programme and costs are maintained, whilst at all times bearing in mind that all legislation affecting the construction of buildings is adhered to. This also includes the health and safety of employees.

i The organisation of the site staff will depend upon various conditions but whoever controls the project, whether it be agent or general foreman, he must at all times work with the clerk of works, to ensure harmony between parties. If all these separate sections are correctly co-ordinated, the project will come to a successful conclusion.

Completion

j Upon the completion of the project, the quantity surveyor should be given all the information he requires to complete the final accounts and to work out extra contractual payments as necessary.

k The contractor must carry out his obligation to complete the defects liability clause, and maintenance work as agreed at the commencement of the project.

Duties and responsibilities of local authorities

These have a responsibility, not only to the client ensuring that construction of the project conforms to the set standards, but also to the public and community at large to enforce the various Acts of Parliament that control the erection, alteration and repair of buildings through the Building Regulations and the Town and Country Planning Acts. It must be remembered that it is an offence to commence construction before approval has been granted even if the building does comply with the necessary Regulations.

Most of the detailed work in relation to the Building Regulations is carried out by the building control officer in whose area the proposed project is to be built. His duties are wide and varied but in outline are:

a Book in plans when deposited, giving register number.

b Check for compliance with Building Regulations.

c Make list of any clarification and/or amendments relating to plans before they are passed.

d May contact architect to discuss amendments. Architect must re-submit amended plans within given time scale otherwise officer will reject application, unless extension of time is asked for.

e Upon satisfactory amendments and passing, officer will issue approval notice. Copy will be sent to architect of plans stamped 'Passed' together with stage construction notices. The architect will pass these notices on to builder.

f The builder is responsible for seeing that the building

control officer receives these notices at specified times, and failure to do so makes the builder liable to a heavy fine, besides which work may be opened up.

Generally notices include:

1 Commencement—states work has started.
2 Excavation of foundations—in correct place and solid bottom.
3 Concrete foundation—checks depth and mix.
4 Damp proof course—laid correctly and at right height.
5 Concrete oversite—correct position and thickness.
6 Drains—open for inspection.
7 Drains—covered and tested.
8 Completion notice sent seven days before occupation or after completion—this is the final inspection before signing it off as being in compliance with Building Regulations.

Besides these statutory inspections the officer may call on site at any time, for spot checks, or when asked to call in to give guidance.

g If during the course of his inspection the officer finds something which does not comply, he may serve notice on the builder to put the work in order. Failure to do so may mean that the local authority will carry out the work and charge the builder the cost of the work in full. The builder does have a right to apply for a relaxation of the Building Regulations.

It is important to the smooth running of the project that a good relationship is struck up between site supervisor and control officer, and discussions should take place over any problems as the officer is there to assist and help under the guidance of the Regulations, satisfying himself that the building is sound and in accordance with the approved plans.

Duties and responsibilities of factory inspector

The factory inspector is employed by the Department of Employment under the Secretary of State, whose task is to ensure that the Health and Safety at Work etc. Act 1974 is implemented (see the chapter on Safety).

Duties and responsibilities of sub-contractors

Basically under the same conditions and responsibilities as the main contractor but overall ensuring that he complies with any agreement between himself and the main contractor because of the necessity and difficulty of co-ordinating sub-contractors into the overall agreed master programme.

Duties and responsibilities of suppliers

a Upon being asked to quote for supplying material, be satisfied that they can complete all obligations regarding any special conditions such as phasing of materials.
b Make sure that samples sent out are a fair reflection of bulk goods.
c Ensure that delivery dates are maintained, issue advice notes as required.
d Give the recognised cash and trade discounts.
e Submit invoices on time for payment.

Nominated sub-contractors and suppliers

The architect may nominate or appoint a sub-contractor or supplier to carry out duties on a project as long as the contractor makes no reasonable objection to their use.

Contracts should be formed between nominated parties and the contractor basically to ensure that materials supplied or work executed is of a satisfactory standard and that for sub-contractors, obligations as to such items as insurances, completion on time, together with small discount allowance on contractor's certificate payment are met, whilst for suppliers, goods showing defects are replaced and, if necessary, contractors compensated for any expenses incurred, also that if payment of invoice from supplier is made within 30 days it is usual for a discount to be made.

3 Legislation

Legislation, the main form of English law, is an Act of Parliament. In most cases, the Government brings in an Act to deal with some national or social need, or possibly to change some existing law. Private members can submit a Bill to introduce legislation but this rarely happens.

A Bill, which in effect is a draft of the proposed Act, goes through a set procedure.

a Introduction of the Bill, to inform all Members of Parliament that a draft exists, if they wish to obtain copies to study—called the first reading.

b At the next reading, points are debated amongst parties.

c If the Bill passes this stage, a Committee, selected from all parties in proportion to their party strength in the House of Commons, discusses any amendments to the various clauses.

d Upon acceptance by the Government of any amendments, the Bill is then passed back to the House of Commons as a Report. Amendments may also be debated at this stage.

e If the Report has not been returned to committee for further consideration, it is given its third reading in the House.

f Upon the passing of the third reading, a similar procedure is adopted by the House of Lords.

Amendments can be made by the Lords, in which case the

Bill is referred back to the Commons to seek necessary approval. As the Commons is the superior House, the Lords cannot reject the Bill, but they may, and often do, delay it for up to a possible period of one year.

g When the Bill has passed through both Houses, Commons and Lords, to become Law it must have the Royal Assent.

h Upon the giving of the Royal Assent, the Commons are notified and the Act becomes law.

Delegated legislation

Often legislation only provides a basis from which to operate. The Health and Safety at Work etc. Act, 1974 is a typical example, where the agency carrying out the details and arrangements of the Act is the Health and Safety Commission, under the responsibility of the Secretary of State for Employment.

Building legislation

1 *Town and Country Planning Acts*
   There are many Acts in existence which affect planning and the control of development, the main aim of them all being to safeguard the public regarding buildings and developments. They are mainly dealt with at present by local government offices.

2 *The Building Regulations, 1985*
   These Regulations are made under the Building Act 1984, bringing together most of the primary legislation affecting buildings. It includes the relevant parts of the Housing and Building Control Act 1985, which lays down the principle changes in the method of Building Control supervision now being done by the local authority or by private approved inspectors, and applies to England and Wales.

3a *Health and Safety at Work Act, 1974*
   This Act has taken the Construction Regulations which formed part of the Factories Act and, together with the

Offices, Shops and Railways Premises Act of 1963, has been brought under the new Act retaining many of the details of the old Acts which can be expected to be modified and updated as time goes on and further safety, health and welfare legislation is required.

3b *The Offices, Shops and Railways Premises Act, 1963*
This is designed to satisfy the safety, health and welfare of persons employed in places of work as suggested in the title of the Act.

Some of the main points with which it deals are:
Cleanliness of premises;
Overcrowding—risk of injury or health;
Temperature—reasonable for working conditions;
Ventilation—adequate, either natural or artificial;
Lighting—suitable and adequate;
Sanitary conveniences—suitable and sufficient;
Washing facilities—conveniently placed;
Drinking water—wholesome and adequate supply;
Accommodation for clothing—provision for hanging and drying;
Seating facilities—opportunities to sit down;
Eating facilities—suitable and sufficient for meals;
Floors passages and stairs.

Other legislation affecting the building industry includes the following:
Building Control Act, 1985
Community Land Act, 1975
Clean Air Acts, 1956–1968
Contracts of Employment Act, 1972
Counter-Inflation Act, 1972, plus subsequent amendments
Defective Premises Act, 1973
Development Land Tax Act, 1976
Electricity and Gas Act, 1968
Employers' Liability (Compulsory Insurance) Act, 1969
Employers' Liability (Defective Equipment) Act, 1969
Employment Protection Act, 1975, 1978
Equal Pay Act, 1970, 1975

Factories Act, 1961 (including provisions for Health, Safety and Welfare on Works of Building and Engineering Construction)

Finance Act, 1971/2 (including Income Tax deduction for sub-contractors)

Guard Dogs Act, 1975

Historic Buildings and Ancient Monuments Act, 1953

Housing Acts, 1964 onwards

Housing (Scotland) Acts, 1944 onwards

Industrial Training Act, 1964

Investment & Building Grants Act, 1971, plus subsequent amendments

National Insurance Act, 1966

Payment of Wages Act, 1960

Public Health Acts, 1936 onwards (complementary to the Building Regulations 1985)

Rating and Valuation Acts, 1925 onwards

Redundancy Payments Act

Restrictive Trade Practices Act, 1968

Road Safety Act, 1967 (Part 2 deals with testing and plating of goods vehicles, design and construction and drivers' licences and inspection of vehicles)

Sex Discrimination Act, 1975

Social Securities Act, 1973

The Employment Act

Trade Union and Labour Relations Act, 1974

Transport Act, 1968 (operators' licences, drivers' hours and records)

Water Acts, 1945 onwards

It is important that supervisors at all levels know about these. It should not be necessary to have to learn and remember them, but three points to bear in mind are:

1 What points do various Acts cover?
2 Where are the Acts to be found?
3 How do you obtain information?

If these points are followed, the possibility of falling foul of the law should be eliminated.

4 Unions

Long before men thought of uniting together as a national body to achieve better working conditions, higher wages, and all the many other things recognised as union activities, small groups of men banded together in the formation of Craft Guilds; these were simply associations of skilled men. Generally to practise a trade, membership of the Guilds was compulsory; in this way a high standard of work was maintained.

Possibly the biggest difference from today's unions was the fact that men accepted conditions as they were, for the aim was then to rise from an apprentice through to journeyman membership and hence to become a master. Bearing in mind whilst a journeyman, he too might one day become a master, any improvements that he demanded and obtained he would be expected to pass on to his future employees, he would put up with things as they were.

With the passing of time and the growth of industry, self-interest began to cause a break-up in the Guilds, with men preferring to leave the old walled towns and seek out new ones springing up free from the authority of the Guilds; these became the real journeymen. The building workers, mostly masons and carpenters, never had much organisation simply because the nature of their craft required their moving from place to place. It was many years from the founding of the first Craft Guilds, believed to be in the eleventh century, to the growth of Trade Unions in the seventeenth century. In the beginning, the Trade Unions had a very hard struggle, not only to organise and obtain legal rights, but even to keep going at all. Between 1799 and 1824, a severe blow was struck against the Unions in the form of the Combination Acts which, in effect, made Unions criminal bodies, with the workers receiving heavy penalties if joining. During this period of legal repression, many Unions went underground. Towards the latter half of the nineteenth century, Union activity increased with local Unions coming together nationally. Unskilled and semi-skilled manual workers were allowed to join Unions; even some of the Craft Guilds allowed this to happen. Many legal battles were fought and Acts passed over a considerable number of years, including the very important Trade Union Act of 1871 which gave Unions a certain legal status, and many more.

Today the major building Union is the Union of Construction, Allied Trades and Technicians, known as U.C.A.T.T., and this acts collectively for its members in the various notes of union activity including membership of the National Joint Council for the Building Industry, one of whose main functions is to produce the National Working Rules.

Employers' associations

These grew with the aim of combating the strength of the trade unions. The function of the employers' associations today is of a collective unit for the purposes of bargaining. They also fill an important function of advising the Government of the day on industrial matters. The employers in the industry have their own federation, the Building Employers Confederation (B.E.C.).

Another association is that of the Federation of Master Builders, which began its existence on 17th July 1941 as the Federation of Greater London Master Builders Limited and was registered under the Companies Act 1929 as a company limited by guarantee.

In 1943 owing to rapid growth the title was changed to the Federation of Master Builders Limited, which acknowledged its national character. 1971 brought the passing of the Industrial Relations Act and as a consequence the F.M.B. broadened its objectives and registered as an employer's association under the Act. Registration was granted and became effective on the last day of 1971, the word 'Limited' being dropped from the title in 1972. Now no longer limited by guarantee, it can take part in activities of a general industrial relations character or, in the F.M.B.'s own words, 'to regulate relations between employers or individual proprietors of that or those descriptions and workers or organisation of workers'.

The National Joint Council

Relations within the building industry between employer and the employee are extremely good. This can be seen by the small amount of time lost through strikes, official or otherwise. The reason for this good relationship is the result of the negotiating machinery that exists to resolve problems and to settle disputes jointly through the National Joint Council, which operates at all levels as shown, Fig. 2.

1 The main functions of the National Joint Council for the building industry are to fix the standard rates of wages of building trade operatives and to determine conditions of employment in the building industry. The Council is established under a joint agreement entered into by the employers and operatives in the industry, which sets out its constitutions, rules and regulations. Relevant extracts from this agreement are given in the 'Explanatory Notes' to National Working Rule 1.

2 The standard rates of wages are fixed by the Council in the manner prescribed in its rules, and official notices setting out the current rates are published by the Council on the occasion of each adjustment.

3 The National Working Rules lay down the principles which govern the terms and conditions of employment in the building industry. It is important to remember, however, that within their framework there are numerous regional or area variations and additions (nationally approved) which need to be ascertained from the Regional Joint Committee

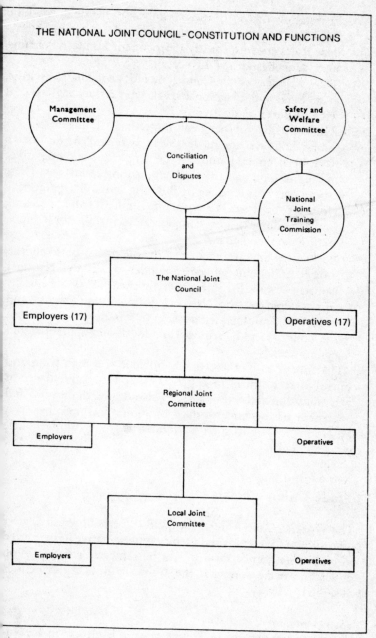

THE NATIONAL JOINT COUNCIL - CONSTITUTION AND FUNCTIONS

Management Committee

Safety and Welfare Committee

Conciliation and Disputes

National Joint Training Commission

The National Joint Council

Employers (17)

Operatives (17)

Regional Joint Committee

Employers

Operatives

Local Joint Committee

Employers

Operatives

Fig. 2

Secretaries or from the appropriate regional or local booklets. In particular there are special Travelling Allowance Rules constitutionally approved for London, Birmingham, Manchester and Liverpool.

4 The machinery of the Council includes conciliation panels, at regional and at national level, whose duty it is to deal with disputes or differences.

5 Arrangements for payment for annual holidays and for public holidays are made under two agreements which have been entered into by the employers and operatives in the building and the civil engineering industries. These agreements are published by Building and Civil Engineering Holidays Scheme Management Limited (Manor Royal, Crawley, Sussex), the company responsible for administering the arrangements.

6 Apprenticeship arrangements in the industry form the subject of a joint agreement under which the National Apprenticeship Scheme is established. This scheme is administered by the National Joint Apprenticeship and Industrial Training Commission (a standing committee of the Council) and by regional and local joint apprenticeship committees.

The functions of these bodies include the regulation and control of recruitment of apprentices, and the surveillance of their education and training. They also deal with disputes and differences arising under a deed of apprenticeship other than those which relate to such matters as wages, working hours, etc.

Trade Union officials

The Working Rule Agreement (W.R.A.) lays down that, after reporting to the employers, a senior representative be permitted to visit site, job or shop so that he can carry out his Union duties and make sure that the Working Rule Agreement is observed.

Site representative
If a site representative or Union steward is elected in accord-

ance with the rules of his union, which can happen only after he/she has been in four weeks' continuous employment with the employer, the Union notifies the employer in writing of the appointment, enabling formal recognition to be obtained by the employer, which cannot be unreasonably withheld or withdrawn.

The W.R.A. sets out clearly the duties and functions of such a site representative besides clearly indicating that only one such official accredited steward being elected for each trade or Union.

Supervisor, Union relations

It is important for all that the relationship between Union and management be harmonious, for it is far better to discuss problems in a cordial atmosphere than to go at it in a heated fashion. The Union official should be given facilities to enable him to talk to the men. His help should be sought if problems occur in obtaining labour. An atmosphere of co-operation and trust will result with this type of liaison, which can only help labour relations.

B.A.T.J.I.C.

Since 30 June 1980 the Federation of Master Builders and the Transport and General Workers Union became parties to the Agreement that the Building and Allied Trades Joint Industrial Council, establishing a body to govern the industrial relations of the adherent bodies forming B.A.T.J.I.C.

The basic objectives of the Agreement being:

1 Form a constitution to devise rules and regulations for the conduct of the council.
2 Determine wages and general working conditions of employees covered by the parties; publish in the form of working rules.
3 Scheme of training for the crafts and skills covered by the Agreement, including apprenticeships registration from 1986.
4 Resolve disputes or agreements arising from the working rules by conciliation panels.

5 Organisation

To meet the growing demand for accurate and competitive tendering which is the main method of obtaining work in the building industry, and to the satisfactory completion of projects, organisation must be the key word. This is done in many and varied ways, but all have one thing in common: the breaking down of the whole into groups or sections each of which will have certain given tasks or objectives, as laid down in the objectives and policy of the firm. This is called an organisation structure. An example is shown in Fig. 3. Each section will have a leader who must ensure that the work of his group is co-ordinated with the efforts of the other sections.

This organisation structure is on formal lines and must be if the objectives are to be carried out. Each section's objectives must be clearly defined in writing so that actions and decisions can be laid at the correct door quickly, enabling answers to be given in a short space of time, when they will be most effective, which can only lead to a better performance all round and reduce the feeling of frustration that can arise through delay.

Although organisation structure is rigid in its concept, informal relationships will develop between the members of the organisation at all levels. This must be allowed to thrive so that a happy team spirit is built up with all the worthwhile benefits to the individual and to the company as a whole that this will bring as long as the formal organisation is never

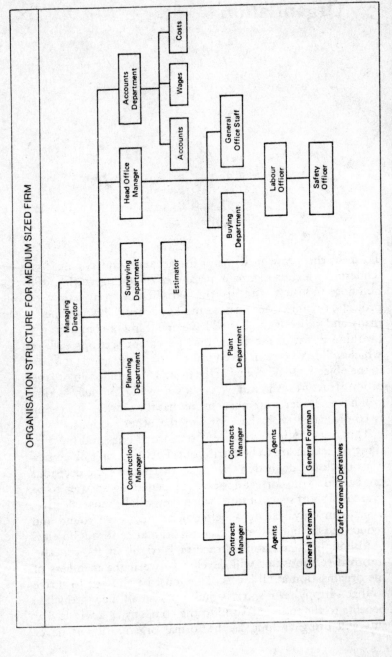

ORGANISATION STRUCTURE FOR MEDIUM SIZED FIRM

Fig. 1

forgotten or misused. In practice, the organisation structure generally will take a shallow or deep form and resemble pyramids when laid down in print.

Fig. 4a, representing a shallow structure, will be found in the thousands of small firms in the industry having up to about twenty-five operatives. This type of structure has the advantage of making communications very easy and quick as the lines between principal and operatives are short and direct. Managerially this is not good practice as it will make the company very vulnerable if something, such as illness, goes wrong at the top, as all the lines of communication will become loose ended with no link between each.

Fig. 4b, showing a 'deep or military structure', is necessary when a bigger set-up working to much finer limits is desirable, requiring more departments or groups for ease of control.

A third form of structure to be found is a combination of shallow and deep structures and is termed the 'line and staff pattern', Fig. 4c. This combines most of the advantages of the two systems with little of the disadvantages.

Span of control

Within the limits of an organisation, the groups, sections or departments will be ruled and guided by a leader who will have the responsibility of ensuring that his section carries out its objectives to a final satisfactory conclusion. The leader, whether he be contracts manager or craft foreman, can only control his subordinates if he can communicate with them. Although the ease of communication will vary under different circumstances, it is generally considered that a practical limit to the number of persons under one's control is between five and seven. With a greater number of persons under one's control, efficiency and production can be affected due to:

a poor communication and co-ordination, which in turn may lead to:

b lowering of morale and the breaking down of team spirit or

c unofficial sub-groups forming within the main group

d lowering of standard of work due to lack of correct supervision

29

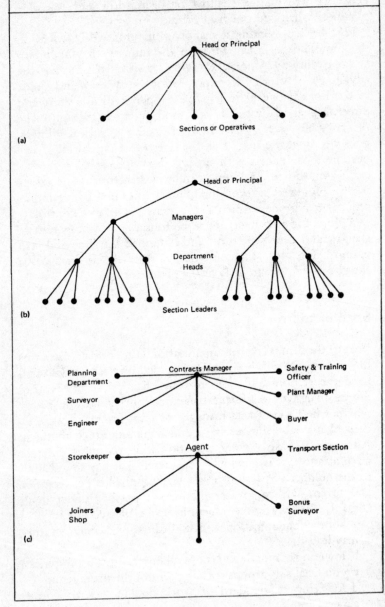

Fig. 4

e poor time keeping and lengthy breaks, and many more unfavourable conditions

As stated previously, the number of persons, five to seven, is in normal circumstances the maximum and when considering a man's span of control all the conditions must be considered bearing in mind such items as:

a the method of communication, e.g. phone, letter, etc.

b the speed in which decisions must be made and their importance

c whether the work is competitive or complex in nature and

d the leader himself considering the load he has to carry in relation to his character and stability.

Example: a joiner's shop making standard joinery: the foreman could be expected to look after more men than a site agent on a half-million-pound project.

A typical span of control for a large size project is shown in Fig. 5, showing which and how many men are answerable to whom.

In any organisation structure or span of control, certain relationships exist. These are expressed as:

Lateral relationship—Persons being on an equal footing, e.g. two general foremen.

Direct relationship—One person able to give an instruction or order that must be carried out, e.g. foreman to operative.

Functional relationship—An advisory nature, e.g. safety officer to foreman.

Staff relationship—No authority but gives assistance generally found in large concerns, e.g. assistant to managing director.

Responsibility and authority

As already stated, any supervisor within the general organisation structure has a specific task; this will be his 'responsibility'. In other words, he is accountable for the success or failure of his allotted task and should receive the awards or penalties resulting from his actions. It is clear, therefore, that in these circumstances responsibilities must be clear, precise and laid down in such a way that each supervisor knows his own duties, thus ensuring a successful completion of the separate units into the final end product of the organisation,

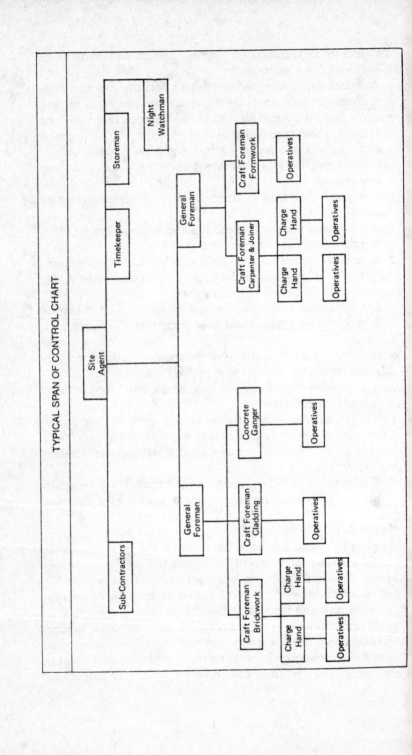

TYPICAL SPAN OF CONTROL CHART

eliminating duplication of work, even the possible failure of work being done.

In general practice, most supervisors pass certain of these responsibilities to their subordinates, this being a good thing for it builds up confidence in the subordinate and motivates him to greater things in the hope, of course, that in time he will receive recognition of his efforts in the form of promotion. It will also benefit the leader of the section or group in that he can put his mind to the more critical and demanding matters of his position. It must, however, be remembered that the leader will still retain full responsibility and be answerable to his superior even for the duties delegated. It is therefore of the utmost importance that with the responsibility to carry out any task a supervisor, leader—call him what you will— must have the 'authority' to carry out his plans, otherwise a feeling of frustration will prevail which will soon show itself in the breaking down of team spirit that any efficient section must endeavour to cultivate in order to achieve its end.

Responsibility and authority must therefore go hand in hand.

Discipline

For any supervisor, discipline is an individual problem. Just as all men are different in their actions and personalities, so the character, temperament and energy of the supervisor will differ. It therefore becomes difficult to generalise. However, there are basic rules that can be followed as a guide to helping to reduce problems in this field. It must first be realised that various stages of disciplinary action are available and will range from helpful suggestions and constructive criticism to severe reprimands and the ultimate and final action—discharge. Which form of action to take depends upon the case in point, but relating the right action in a given situation is a major aspect in the handling of people.

General rules or principles to bear in mind when disciplining anyone are:

1 Always be sure of the facts and that you have the right person to be reprimanded.

33

2   No favouritism or discrimination should be shown.
3   Do not lose your temper, even though this may be difficult; keep your self-control and stay calm.
4   Always discipline in private and not in the presence of others, whether superiors or subordinates.
5   Be fair—but firm.
6   Decisions and behaviour must be consistent with the offences.
7   Be prepared to let people present their side of the story.
8   If the error is on your side, admit it.
9   Do not turn your back on disciplinary problems.
10 Never harp on a misdemeanour or bear malice.

Without a high cultivation of discipline, a slack anti-co-operative situation will arise, resulting in lower production, poor teamwork and a general loss of control.

Leadership

Anyone in a position of commanding, controlling or co-ordinating men requires the 'know-how' of how to get the best out of them, and this ability or skill is generally referred to as leadership. Many men in positions of authority are said to be born leaders and this is often true, for men with this attribute will emerge in any group, even if a leader has not been appointed, mainly through imposing a strong personality. This leadership is necessary in all walks of life, for without it there would be chaos, with everyone telling each other what to do and no one doing any work. Could you imagine a country without a government or dictator, or an army without a commander? It must not be thought, however, that leaders are always born to fill these positions of authority; many excellent leaders have emerged *acquiring* the art of leadership through experience, training and hard work. Training plays its part in the development of a leader, for skills can be given to the potential leader that will give self-confidence, a characteristic a leader must have. There is no clear-cut way of spotting on sight, a leader in any group or gang and as already stated, some may emerge but generally they are recognised by their personal qualities; these show themselves over a period of time. The following are some of the qualities one could expect to find in a leader:

Self-confidence: of great importance; without it the men under him will soon lack confidence in his orders, for the leader will be continually changing his mind because he is not sure of his own knowledge. A leader must be positive in his thinking.

Temperament: a quick or erratic temper will soon lose the respect of the men. Temper should be even and controlled; a cheerful enthusiasm for work with consideration and fair play for others will contribute a great deal towards success.

Respect: a leader needs the respect of all those with whom he is working, not only subordinates but also superiors, and this popularity can often be achieved through consideration for others. Bad manners and language will do nothing to enhance respect.

Tact: many problems in labour relations present themselves to a leader and so if unpleasantness is to be avoided and good feeling maintained, situations and unpleasant tasks must be handled diplomatically.

Energy: example is the keyword here, for it is generally agreed that enthusiasm is infectious; spurts of energy are no good, it must be sustained, will often show itself in the organisation of a site.

Technical skill: comparisons are often made on this subject relating to skill gained through years in an industry and that gained through education and books. The best solution would be comprehension of both so that the leader knows the methods suggested in theory and can relate these through his acquired knowledge of practical application. Some 'book learning' should always be sought by the leader for it must be realised that points picked up in practice may not always be the right ones.

There are many other attributes we may look for in a leader, e.g. tidiness, punctuality, intelligence, integrity. To find all of these attributes in one human being is rare, but some can be cultivated if the basic qualities are there, and, possibly above all else, the person chosen must *want* to be a leader, whether in the capacity of craft foreman or managing director.

Team spirit

Smells like [handwritten annotation above "Team"]

The building industry relies a great deal upon the co-ordination and teamwork of relatively small groups. If any project is to be successful, not only in making a profit but also to finishing by the specified completion date, all concerned will have to work together in goodwill and harmony. Team spirit depends to a very large extent upon the ability of the leader, ensuring that the feelings, personalities, likes and dislikes of all the parties in a team are welded into one efficient working unit, a very difficult task requiring a great deal of skill.

There are many ways in which to encourage people to work. It is therefore most important that a leader knows his team. Men must be treated as individuals and not just as a number or a machine. The supervisor who is prepared to spare a few minutes from his duties for a chat with an operative, finding out a little of his background, family, hobbies and general character, will soon reap the rewards when having to deal with the team as their leader. He will have assessed how each and every man must be treated. This does not mean that the supervisor should allow familiarity to cloud his judgement; it is worth remembering that familiarity breeds contempt. His approach must always be firm but fair, whilst at all times he must be prepared to listen to grievances.

Motivation

The art of getting someone to *want* to work is called 'motivation' and is closely akin to team spirit; for the leader's aim is to get people to work, and work together. One of the first aims is to try to instil a feeling of belonging. This can often be achieved by informing the team as to the progress of firm or site; this will help encourage a sense of participation which in its turn tends to create a feeling of keenness and loyalty. The supervisor should also try to ensure that a workman's own initiative is allowed sufficient scope to prevent him from becoming frustrated. He should be encouraged to seek promotion—no good leader stands in the way of anyone's advancement.

There are many ways in which a supervisor can obtain this general good and productive feeling, and it is generally up to the individual how it is achieved, but whatever method is adopted, it must be realised that the economical rewards must be a major consideration. It is therefore necessary that a sound wages policy is laid down with perhaps the added incentive of an attractive bonus scheme. Other considerations to aid motivation include: security, status, welfare, sound opportunities.

Decisions

If a single course of action is clear, due to having all the requisite information, an automatic decision can be made. Unfortunately, due to many variables that often occur this is not usually the case, especially in the construction industry.

The big problem with decision making is that it is a human situation and so whatever the processes to be used or whatever information is available the final decision will be related to the thinking of an individual.

Other problems may of course bear a relationship to the decision made. For example, the speed at which the decision has to be made, which in many cases may affect the evaluation of variables and the resulting possible courses of action.

Monetary decisions are often critical on construction sites, especially when margins are small, requiring as much objective consideration as possible to prevent losses.

Before coming to a decision the manager should:

a evaluate the problem by breaking it down into recognisable elements

b analyse the elements using as much data and information that is available, examining the likely outcome of each

37

c look at the various alternatives or outcomes and make what
is a subjective assessment of each

When these issues have been carefully studied it is more likely
that by choosing from among the various options considered a
satisfactory decision on the problem as a whole may be
determined.

The complexity and importance of decision making generally
relates to an individual's position in the organisation structure
and the responsibility a person has, but whatever the position
held and whatever technique is used, including the much wider
use of computers in this field, the final decision is still that of the
individual.

6 Personnel

The building industry, although becoming more mechanised as years go by, still relies a great deal on the efficiency of manpower. It is necessary, therefore, that all supervisory and management staff should have a good understanding of the various activities related to personnel. Some aspects of personnel management have been dealt with in the relevant chapters.

Recruitment of staff

Every employer wishes to employ only the best, most efficient and enthusiastic staff available. He requires this type of manpower to ensure the prosperity of his organisation.

It is therefore of considerable concern to all that only the right people are recruited, and this is a very difficult task at any level, relying very much on chance. However, to reduce the risk of poor or incorrect selection, certain procedures can be followed which it is hoped will eliminate error. If an organisation has a personnel manager this is of considerable help, as it is expected that he would, through his knowledge, reduce the risks of poor selection. When new staff are required, heads of departments or sections should notify the personnel manager (or whoever is responsible for recruitment of staff), clearly indicating as much information as is available as to requirements, taking into account such details as sex, age, qualifications preferred and the type of work, etc. It is then the personnel manager's problem to try and fulfil these requirements.

Appointment services

There will of course be various methods of obtaining staff, dependent upon the position on hand.

Trainees/apprentices: when young people are required, the local Careers Officer may be able to advise. Local schools or technical colleges who have young people on full-time courses looking for employment may also be able to help.

Skilled and unskilled operatives: operatives can generally be obtained through the local employment offices or by advertising in the local press.

Professional and management staff: can be obtained through professional institutions, universities, or through the local employment office by seeking those on the professional and executive register.

Advertising: is a very attractive method of recruitment, especially for management posts. By using the national press or trade publications, although relatively expensive, the net can be cast over a larger area, hopefully catching the right calibre of applicant. A typical advertisement is illustrated opposite, highlighting the areas of importance to attract the right type of applicant.

Application forms

Upon receipt of an enquiry, an application form may be sent to the applicant, requesting details of his background. This is not always so, especially in the case of smaller concerns where a letter applying for a post in the applicant's own hand will satisfy.

Application forms have the advantage that they give a company the information it requires in an easily defined way, giving a reasonable assessment for initial selection of candidates.

These application forms are divided into sections, depending on the information the company requires. For example, personnel particulars, general education, further education and training, employment history, medical history, name and address of referees. Space is often left for applicants to write a

ESTIMATOR/SURVEYOR	Job Title
YORK Area	Location
£9,300–£10,100 per annum	Remuneration
To be responsible for all estimating within the small works department of a medium-sized building company. The post holder will probably be in his/her mid-thirties, with an H.C. in Building or Surveying, and preferably corporate membership of a professional institution.	Job Description Qualifications
Applicants should have experience in both fields of estimating and surveying in a small works department.	Experience
Contributory superannuation scheme, casual user car allowance. Removal and subsistence allowances in approved cases.	Benefits

Application forms, which should be returned by 20th October 1987, can be obtained, together with further details from: The Personnel Manager,
      Messrs Cold & Snow,
      257 Bleak Street,
      Northington,
      County LYK 3LW
      Telephone: Northington 12345
            Ext 67
	Closing Date
	
	Address of Company
	
	Telephone

letter of application. This also is an indication of writing ability and use of English, and often gives an insight into an applicant's enthusiasm for the post.

Application forms are then critically examined and it is hoped they will produce suitable candidates, the number selected varying considerably with the type, position and requirements of the post.

Short list

The number of applications selected from the application forms make up a short list of candidates. This implies that these are the most suitable ones from the application forms and are therefore considered worth interviewing for further selection. A letter is sent to those selected, advising them of the time and date for the next phase of selection. During this period of time the candidate's references are taken up from the referees named on the application form, usually 2 or 3 in number.

References

These are used in the hope of finding out information that may have been concealed from the application form. They should always be treated with certain reservation as an applicant will only ask those who think quite highly of him to act as referee, so the truth may be a little clouded.

Conducting the interview

This is a task requiring skill and careful preparation. It is essential that the interviewer knows precisely what is wanted and what questions to ask, enabling him to obtain all the necessary information the firm requires, dealing in full with matters on the application form and matters not dealt with adequately on the form. Considerable thought must be given by the interviewer to the questions to be asked, so that a 'yes' or 'no' answer is not the result. The personal qualities of the applicant must be carefully assessed, bearing in mind always the type of work the post entails. For example, a salesman will most likely present a different personality than that of a senior estimator. Often the preparation of a simple document listing points to be assessed will be of advantage. This is especially useful if a number of applicants are being interviewed at one time, because it is difficult to remember each person clearly and relate information gleaned back to the right person. A list of assessment points applicable for the post, with some form of grading system, such as excellent, good, average, poor, or even fractions $\frac{1}{5}$–$\frac{4}{5}$, should be indicated against the word while the interview is in progress. These can easily be assessed on completion of the interview and may be of considerable help in recalling applicants to mind when a decision has to be made.

The interview: should be held in pleasant surroundings. The interviewer should be informal and friendly. This has the effect of making the applicant feel at ease and talk freely and naturally – a very difficult objective to obtain in normal interview situations, thus making it hard to get to the personality of the interviewee, especially in the short time generally allotted for this task.

The interviewee: must try to make a good impression. This can be achieved if the following are considered:
1 Ensure appearance is smart
2 Speak and act naturally
3 Keep to the point
4 Even if sitting, maintain a good posture
5 Do not rush to answer questions on background, hobbies, education, etc.

At the end of the interview, you may be asked if you have any questions. If you have, ask them, for generally this shows that you are interested in the post. On leaving the interview, a 'thank you' to the interviewer will do no harm.

Testimonials: you may be asked to take along copies of testimonials, although many firms have now discontinued this practice. If you have qualifications, these should be taken to the interview to substantiate claims of awards.

Final selection

This is made by the interviewer(s) considering all relevant facts, application form, interview, assessment sheet and other additional material, such as work folders, projects, portfolios, etc. The successful applicant may be informed at the end of the interview period, or in writing at a later date. If an applicant is given verbal notification, this should always be confirmed by letter. It is then the applicant's place to accept (or reject) the position offered, remembering that he must consider all the implications of a move.

Commencement

Upon commencement with a new organisation, the new employee should be informed of the company's rules and all other items of general information, including such things as:

Pay – basic overtime, bonus, etc.	Disciplinary procedures
Breaks	Timekeeping and recording
Holiday qualifications	Report of absence
Pension schemes	Fire drill
Education and training	Accidents, etc.

Contract of Employment

Employee's individual rights have been considerably improved by legislation during the past 20 years with provision mainly contained in the Employment (Consideration) Act 1978, as amended by the Employment Acts 1980 and 1982. Besides these Acts other individual rights are also covered in the Equal Pay Act 1970.

These Acts give most employees in Great Britain the right to a minimum period of notice of termination of their employment, according to length of service and the right to receive from their employer a written statement of their main terms and conditions of employment and certain other information.

Items to be noted in a typical contract are given below:

1 Name of employer and employee
2 Interval and scale of remuneration
3 Normal hours of work
4 Holidays and holiday pay
5 Terms and conditions relating to injury and sick pay
6 Pensions and pension schemes
7 The length of notice of termination by (a) employer
(b) employees
8 Date of commencement of employment, including job title
9 Signature

All employees should receive such a contract of employment not later than 13 weeks after commencement of work.

The Employment Protection Act

The purpose of this Act is to promote the improvement of industrial relations through the Advisory Conciliation and Arbitration Service and encourage the extension of collective bargaining; together with these two items it gives employees new rights and provides for greater job security.

The Act covers all full-time and a large number of part-time employees, it also gives Trade Union members new rights. Rights to recognition and information and consultation in some areas are established. Briefly the rights for employers are:

1 *Guarantee payments*
   If a worker is put on short time or is laid off, and a full day's work is lost, the worker shall receive a normal day's pay with a maximum limit of £x per day, pay being guaranteed for five days in any three-month period not a fixed calender quarter.

2 *Medical suspension*
   If a worker has been suspended from work under statutory regulations following examination by an appointed doctor or employment medical adviser, he will be paid normal wages during suspension for a total of 26 weeks.

3 *Maternity*
   For the first time pregnant employees are protected by law and can no longer be dismissed. For the first six weeks of absence the employee is entitled to payment of wages at the rate of nine-tenths less flat rate National Insurance maternity allowance.

4 *Trade Union membership*
   An employee may join an independent Trade Union and cannot be victimised by his employer and can take part in Union activities.

5 *Time off*
   An employee is entitled to time off for:
   a Trade Union duties and activities
   b public duties—Justice of the Peace, etc.
   c looking for work

6 *Insolvency*
   So that an employer will not lose money if the firm becomes insolvent, the Department of Employment may pay outstanding debts relating to a period of up to eight

weeks and a maximum of £155 a week per employee. (1987)

7   *Periods of notice*
One week's notice after four weeks' service, two weeks' after two years' service, and thereafter an additional week for each year of service to a maximum of 12 weeks after 12 years.

8   *Written statement of reasons for dismissal*
Employee must ask for written notice.

9   *Remedies for unfair dismissal*
May seek reinstatement or re-engagement or may be awarded financial reward.

10 *Redundancies*
Employer must consult appropriate Trade Union.

Redundancy

The Statutory Redundancy Payment scheme is administrated under Part VI of the Employment Protection (Consolidation) Act 1978 as amended by Part III of the Wages Act 1986.

A payment is made to any employee who has worked continuously in the service of an employer, for not less than 104 weeks since the age of 18, normally working over 16 hours a week, and is due to be dismissed by the employer.

Payment can also be due if an employee is laid off, receiving no wages, or put on short time, receiving less than half a week's pay, for four consecutive weeks, or six weeks in a continuous period of 13 weeks. In these two cases an employee may claim, in writing, for redundancy payment without waiting to be made redundant.

However the employer can if he/she believes that normal working arrangements will return within four weeks, resist the redundancy request.

Lump sum payments for employee redundancy, will be calculated from completed year of service up to a maximum of 20 years as follows:

1   For each year of employment between 18 and 21 inclusive, half a week's pay.

2   For each year of employment between 22 and 40 inclusive, one week's pay.

3   For each year of employment between 41 and under 65 (60 for women) inclusive, one and a half weeks' pay.

The limit used in the calculation of 'a week's pay' is £158 (1987) being adjusted periodically.

Employers

Employers are required to consult appropriate trade unions about proposed redundancies, even if the firm seeks voluntary redundancy. Minimum periods of consultation are laid down as: 10 to 99 employees may be dismissed with at least 30 days notice; 100 or more with at least 90 days. Redundancy rebates have now been abolished, so that the employer has to bear the whole cost of making employees redundant if they have ten or more employees.

Holidays with pay

In 1942 both sides of the industry signed an agreement to provide operatives with annual holidays with pay. This scheme, run by the Building and Civil Engineering Holidays Scheme Management Ltd., now provides all employees in the industry with paid annual holidays. Because of the great movement of labour in the construction industry, this special method of paying employees for holidays was evolved. The basic points are as follows:

1   Each employee has one card in three parts to provide annual holidays.
2   Employers purchase holiday stamps in block from the management company (also cards when required). Employees do not contribute to the scheme.
3   The first employer in the week, commencing Sunday midnight, affixes a stamp to each card.
4   Where an operative is absent due to sickness, provision is still made for him/her to receive stamps. This entitlement allows for a maximum of eight to be given, provided that the operative worked five days and the whole period is covered by certificates of incapacity to the employer's satisfaction.
5   To receive payment from the management company, the employer sends the receipted cards and prepared reimbursement form, completed and signed; he will then receive back

the amount as shown in stamp face value. Care must be taken that the registration certificate issued by the Post Office when the cards are sent in is retained, as the onus of proof of posting rests with the employer if they are mislaid.

Holidays Under the National Working Rule Agreement provides that each operative (including apprentices/trainees) have a holiday entitlement of:

a seven working days in conjunction with Christmas, Boxing and New Year Days, giving a two week period

b at Easter four working days immediately following Easter Monday giving a one week with summer holiday of two weeks, which must be taken between 1 April and 30 September

c one further public holiday is granted for May/National day.

Apprentices: To ensure that apprentices coming out of their time receive the recognised holidays with pay, the employer starts a card for the apprentice in his last year.

Death Benefit Scheme

Administered by Holidays Scheme Management Ltd, it is designed to provide a cash benefit to dependant(s) on the decease of an operative. Employers are required to operate this scheme as laid down in the Working Rule Agreements. As with the Holiday Scheme the employer meets the full cost of the scheme for those over 18 years, up to 65 years of age.

The scheme requires the operative covered by the scheme to have completed at least four weeks of reckonable service before dependant(s) may claim benefit. The maximum amount of this benefit is amended from time to time by the Building and Civil Engineering Joint Board. In 1985 it was £4500; it is calculated as follows:

No of weeks of Reckonable Service at the date of death	Maximum Lump Sum Benefit
Less than 4	NIL
4 to 9	25% of the maximum benefit at the date of death

10 to 19	50% of the maximum benefit at the date of death
20 to 39	75% of the maximum benefit at the date of death
40 or over	100% of the maximum benefit at the date of death

Death in Service

This allows for higher maximum amounts to be made, in 1985 this being £9000, and covers not only death occurring as a result of an accident at the place of work in the construction industry but also travelling to or from such place of work.

Reckonable Service

The schemes are taken as being from 5 April 1982 onwards, denoted by adult stamps affixed to holiday cards. No tax is payable on these awards and in the case of death in service they do not form part of the operative's estate. Payments are at the absolute discretion of the Trustees who consider the operative's circumstance at the date of death, and the weeks of reckonable service.

Employed

If the operative's employment was terminated by the employer, cover through death benefit is automatically continued for four weeks of subsequent certified employment; after this period any payment is made at the absolute discretion of the Trustees.

Operatives

A person with 200 or more week's Reckonable Service may have benefits payable regardless of operative's circumstances at the date of death, benefit being calculated as number of weeks Reckonable Service × Retirement Benefit rate.

Salaries and wages

A salaried person is generally a member of the firm's permanent staff, and is paid at regular intervals, usually at the end of each calendar month. There is little variation in the amount paid; the salary is simply calculated as one-twelfth of the year's gross pay, so unless the deductions, such as income tax, National Insurance, etc. alter, the basic monthly amount should be the same.

Wages, usually paid weekly, are more difficult to calculate, especially in the building industry. In case of repair work, men may move to three, four, or even more jobs in one week; and of course, there is always the problem of loss of earnings due to bad weather; travelling time may have to be paid, or lodging allowances, and many more such items. It is therefore vital that careful records are kept on workmen's time to assess correctly their earnings.

Equal pay

The first Act covering equal pay was in 1970. This has since been supplemented by the Sex Discrimination Act 1975, the Employment Protection Act 1975 and the Equal Pay (Amendments) Regulations 1983 and as the title suggests the Acts are designed to eliminate discrimination between men and women in pay and other terms of contracts of employment, e.g. bonuses, holidays, sick leave, etc.

Many disputes have and are still occurring on this matter and a woman who believes she has the equal rights of a man, which the employer will not accept, may apply to an Industrial Tribunal for a decision; as with unfair dismissal, conciliation is also used to promote a settlement.

Sex discrimination

This Act came into force in 1975 making it unlawful to treat anyone, on the grounds of sex, less favourably than a person of the opposite sex.

The Equal Opportunities Commission was set up to ensure the effective enforcement of the Sex Discrimination and the Equal Pay Act, promoting equal opportunities between the sexes.

The Commission provides for investigation recommendations

to Government on existing law, and provides help to individuals making complaints.

Under the Sex Discrimination Act 1986, small firms (which form a large part of the construction industry) with five or fewer employees were exempt from the 1975 Act until 7 February 1987; they now have to comply with its employment provision.

The Act also provides that employers will no longer be able to set different compulsory retirement ages for men and women.

Clocking on and off

This usually falls into one of three categories:
1 The time-clock
2 The tally method
3 The time sheet or book

Time-clock: each employee will have a time-card which, upon arrival or departure from his place of work, he will insert into the time-clock which automatically records the time. Many of these clocks will, after a certain period of time, record in a different coloured ink, generally red, to signify lateness. After, say, a quarter of an hour, the clock should be locked so that the late-comers must report to the foreman or time-keeper to explain the reason for lateness. The advantage of the system is that accurate times are recorded without supervision, although this can lead to men clocking on and off for each other, a practice that must be guarded against. The other main disadvantage is that it can cause wasted time through congestion round the clock.

Tally board: simply a 'key' board with a tally on each hook. Every operative has a tally bearing his own name or check number. When he arrives each morning, it is either turned over or placed on the reverse side of the board. This shows clearly which men have arrived, but does not give the time. It is therefore necessary, after clocking on time, to have someone to record the period of lateness. The main advantage with this system is that a quick check can be made on arrivals or departures.

51

A.N. OTHER Ltd.
REDHILL

TIME SHEET

NAME ....

CHECK No. ....

TRADE ....

WEEK ENDING ....

Job No	Job	Description of work		MON	TUE	WED	THUR	FRI	SAT	SUN	TOTAL	RATE	TOTAL	RATE	TOTAL	RATE	TOTAL
			Plain Hours														
			Bonus Hours														
			Plain Hours														
			Bonus Hours														
			Plain Hours														
			Bonus Hours														
			Plain Hours														
			Bonus Hours														
			Plain Hours														
			Bonus Hours														
			Plain Hours														
			Bonus Hours														
			Plain Hours														
			Bonus Hours														

FARES
EXPENSES
SUBSISTENCE
TOOL MONEY

TOTAL WAGES £

TOTAL BONUS £

TOTAL EXPENSES £

TOTAL £

GROSS ENTITLEMENT £

A.N. OTHER Ltd/AGENT ....

AUTHORISED ....

WEEKLY TIME SHEET

SITE ....

WEEK ENDING ....

Name and Initials	Trade	Mon	Tue	Wed	Thur	Fri	Sat	Sun	Total	Rate	Basic	Bonus	T.A.	Subs	Expenses	Remarks
									£							

Fig. 7

Time sheets: used mainly on small jobs when a foreman fills in a time sheet, an example of which is shown, Fig. 6, for each man, or by the individual if on repair work on his own, or similar.

Whatever method of recording is used, the head office wages department will require details. This is usually submitted on a weekly time collection sheet, Fig. 7. The wages department will then transfer the relevant information on to an employees' individual pay record.

Deductions

When the employee's gross wages have been calculated, general deductions have to be made in the same way as a salaried person.

Income Tax: generally the biggest single deduction from most people's wages. Tax must be deducted by the employer at the rates specified by the Inland Revenue Department, in relation to the code number issued to the individual by the tax department. This code number is worked out by a rather complicated system of allowances which, when totalled, give the amount of tax-free earning possible; all other earnings are taxable, e.g. salary, wages, bonus, overtime, etc.

National Insurance: everyone has a weekly contribution deducted at a standard rate if *not* contracted out of the new N.I. scheme which came into effect April 1978; or if contracted out a fixed rate is used with additional percentage plus rates deducted related to earnings.

The yearly tax deduction card P.11 is used to record both employer's and employee's contributions.

Payment of wages

Employees may, under the Payment of Wages Act, 1960, receive their payment in one of the following ways; other than by the standard method of a wage packet:
1   Direct payment into employee's own bank account
2   By cheque

3 By money order or
4 Giro cheque payable at post offices

Manual workers are required to put in writing their agreement to the method adopted; salaried staff have no option but must accept the company's method of payment. The advantages of these methods of payment over that of paying in cash are:

    a Wages staff is reduced as the task of packeting cash is abolished
    b Less security is required to guard against theft, not only at head office, but also in fetching and carrying

The practice still adopted by the majority of builders is of paying out, using a pay packet. Although various types exist, most now enable the employee to check contents without opening the packet. A breakdown of the wages must also be shown, either in the form of a slip or as part of the pay envelope.

Itemised pay statement: under the Payment of Wages Act only certain categories of manual workers had a right to receive an itemised pay statement and this only if the worker himself asked to be paid otherwise than in cash and the employer complied with the request. The Employment Protection Act, with certain exceptions, states that every employee will be entitled to a detailed pay statement which shows gross pay, variable and fixed deductions and net pay. The employer may issue a standing statement of fixed deductions but only if any alterations are notified to the employees when they are made, and it is re-issued at least once a year.

Termination of employment

The Working Rule Agreement allows for notice by either employer or employee to be given as follows:

1 During the first five normal working days of employment:	Two hours' notice to expire at the end of normal working hours on any day
2 After the first five normal working days of employment:	One clear day's notice to expire at the end of normal working hours on a Friday

3	One calendar month's continuous employment, or more, but less than two years:	One week's notice
4	Two years' continuous employment or more but less than twelve years:	One week's notice for each full year of continuous employment
5	Twelve years' continuous employment or more:	Twelve weeks' notice

Termination can be carried out by:

  a mutual consent (in writing)

  b payment of wages (at flat rates) in lieu of prescribed period of notice

  c misconduct

  d stoppage of work by recognised competent authority (one clear day's notice to end at normal working day's end)

The Employment Protection (Consolidation) Act 1978 as amended by the Employment Act 1982 also gives provision to rights of notice and reasons for dismissal.

Unfair dismissal

This will be heard by an industrial tribunal, available to most employees who feel they have a complaint regarding unfair dismissal (employees being recognised as someone who works or worked under a contract of employment, which may be written, oral, or implied). To be able to act on unfair dismissal an employee must have completed not less than two year's continuous employment with their employer; or at least eight hours a week for at least five years.

Typical items

Items subject to possible unfair dismissal would be, for example, if an employee was dismissed on account of trade union membership or activities because of non-membership of a trade union. Or an employee may have been dismissed because of pregnancy or reasons connected with pregnancy, unless it is impossible to carry out the job she is doing adequately. If after pregnancy her employer fails to take the woman on, after maternity absence, this may also be a case of unfair dismissal.

Conciliation
This can be attempted in the cases mentioned above, and other such like cases, at the request of either party, by a conciliation officer who after receiving a copy of the employee's application and considering there is a reasonable chance of success, will take steps to see that parties reach a voluntary settlement.

Tribunal hearing
If a settlement is not reached or withdrawn a full hearing of the tribunal is held. The tribunal consists of a legally qualified chairman and two lay members, and is kept simple so that applicants may put their own cases, although they can be represented by a friend, trade union officials or a lawyer.

Compensation
If the finding of the tribunal goes against the employer, who makes no order to reinstatement or re-engagement of the employee, alternative remedies of compensation are awarded usually taking the form of a:
1 Basic award. This is based on employee's age, length of service and weekly pay, being calculated in the same way as are redundancy payments; or employee may receive a:
2 Compensatory award, where the tribunal considers what is a just and equitable amount for the loss the employed has suffered owing to dismissal; a sum subject to periodic renew, giving a maximum figure.

Upon leaving employment, the operative is given the following documents:

Upon leaving employment
The operative is given the following documents:
1 P.45 Income Tax form showing the total amount of pay and tax paid to date since commencement of tax year
2 Holidays with Pay cards.

The employee should be made to sign for these documents. New employees must produce these documents at commencement of work with new employers.

7 Tendering arrangements

The majority of builders still obtain much of their work by the system of tendering, especially for new work or work of some importance, as opposed to smaller or repair work where the client will generally approach the local builder—or a recommended one—to carry out the work.

Tendering takes various forms but all have one thing in common—they cost money to prepare and when it is considered that any number between four and twelve firms could be after the same work and that only one will be lucky, it is clear that much money is wasted by the remainder. It must also be remembered that on average a builder will only be successful on approximately three out of every ten tenders submitted. The money lost on the seven unsuccessful ones will have to be made up on future tenders or written off as a loss.

Open tendering

Many local authorities, particularly the smaller ones, still favour this method. An advertisement is placed in the national and/or technical press asking for tenders for construction work or often the supply of materials; upon receipt of tenders, the lowest sum will be selected.

To prevent anyone from sending for particulars just to

'have a look', it is a general rule for the advertisement to state that a deposit of up to about five pounds must be paid, which is returned only upon receipt of a bona-fide tender. This system of tendering has been condemned in various reports and when note is taken of the disadvantages of this system it is clear why.

Disadvantages: much time is taken up and lost through the placing of notices in the press, the preparation of the drawings, bills, sending documents out and receiving the same volume back to be sorted and checked. The lowest tenderer may not have the resources, manpower or management to carry out the project to a satisfactory conclusion, and so consequently will not be a suitable contender for the work. Poor workmanship and use of unsatisfactory materials to combat low cost figures are seldom justifiable and this can result in the use of an 'unknown' contractor. The only safeguard is that the lowest tender need not be accepted, but if public money is being spent, it generally is.

Selective tendering

The main advantage of this type of tendering system is that all those asked to submit a tender will be capable of completing the project in a satisfactory manner, because the resources and management organisation will be known: the whole reason for selecting a particular firm.

How are they selected? Generally in one of two ways:
1 Builders of known experience and reputation whom it can be reasonably certain will carry out the work to the required standards are asked to tender.
2 By drawing up a short list from applicants who have been given brief details of a project, again through the national and technical press. It can be seen that there is a similarity here in that the method of seeking out potential tenderers is the same as in the open tendering system, the difference being, of course, in the drawing up of a short list of builders of known capabilities.
3 Local authorities and architects in private practice have a

list of builders who have done work of a similar nature satisfactorily in the past. It is usual for these lists to be revised annually.

A general guide to how many builders should be asked to tender is shown below as indicated in the Code of Procedure for Selective Tendering 1977.

Value of project	Number of firms to tender	
up to £50 000	5	
£50 000–£250 000	6	amended
£250 000–£1 million	8	periodically
£1 million	6	

Five and eight are considered the minimum and maximum numbers, with the provision of one or two names to act as reserve for any firm not wishing to accept the invitation to tender.

Disadvantages to the system:
1   Builders knowing other members on the list may work to their own rota system to share out work, leading to high prices.
2   Unless lists are amended at frequent intervals, firms of improved ability may not be considered.
3   Lists are still often too long—this may be advantageous if list is rotated so all have an equal chance of work over a period.

This method, although having faults, will in general reduce costs and save time.

Negotiated contracts

In this system of tendering, the client will leave much of the negotiating to his professional advisers who, upon choosing a builder capable of carrying out the work, will resolve terms.

The main reason for this type of tendering is either:

a that the project under consideration is to be completed as early as possible with no delay, as in normal tendering procedure, in the preparation and despatch of documents and selection of a builder, or

b that much of the work to be carried out is of an unknown nature, resulting in difficulties of detailing, billing and costing.

The two reasons mentioned need not be the only ones for negotiating a contract: a client may wish to retain the services of a builder who has proved reliable by giving complete satisfaction in the past. There may also be situations that are best resolved by bringing a builder in right at the earliest discussion stage. In this form of tendering, great care must be taken in the formulation of the general contract agreement regarding costs. One of three methods is used to calculate a final project price:

Cost plus contract: In this method, all labour, sub-contractors, materials, plant, etc., are carefully recorded and documented and to these recorded totals the rates that have been negotiated are added, so giving a final total cost. For example, if we assume the negotiated plus rate for labour was 20% and the record labour costs paid out to the operatives by the builder were £16 000, the amount to be added to the final account for labour would be: £16 000+£3200=£19 200. A good relationship between parties must exist to achieve satisfactory results in this type of contract as there is little incentive for the builder to keep costs down. The higher the cost, the bigger the builder's return.

Cost plus fixed fee: A fee quoted in pounds is inserted in the contract instead of the plus percentages as in the cost plus contract. For this method, a fairly accurate estimate of costs must be worked out so that the negotiated fee can be realistic and satisfactory both to the builder and to the client. Again, as in the cost plus contract there is little incentive for the builder to economise as the fee remains fixed.

Target cost with variable fee: This is similar to the other two previous negotiated tenders, the big difference being that the fee or profit is fixed neither by a percentage nor a set sum, but with a fee in pounds that can be increased or decreased by means of an agreed percentage between the parties.

A contract of this type will provide an incentive to a builder to complete his works at as low a price as possible, for in so doing he will benefit by receiving the agreed percentage of the savings. Alternatively, upon the project costing more than the agreed target cost, he will forfeit some of his fee.

Example: The target cost of a project was estimated to be £60 000; the percentage plus or minus rate was agreed at 25% with a fee of £5000. If on project completion the cost totalled £62 000, the builder's fee would be reduced by 25% of £2000, the amount over the target figure. The fee received would be £5000 − £500 = £4500.

It can be clearly seen that a reasonably accurate estimate of the work would need to be worked out before the builder committed himself to the variable fee.

Advantages of negotiated tenders are:

1 A builder will be brought in at a very early stage for consultation with the designer.
2 Time and money are saved on the work done in unnecessary tendering.

Against this system is the high degree of mutual trust that must exist between the parties concerned with the project, and, business being what it is, this is not always found.

Serial contracting

This type of contract could be classed as a negotiated contract. A builder is asked to tender for one particular project, bearing in mind that if he is successful he will be asked to negotiate for a number of projects of the same character. For example, a local authority intending to build four or five schools of the same type within their area.

With this type of contract, accurate bills of quantities and a careful estimate can be prepared, the builder bearing in mind that if he succeeds in winning the contract he is assured of a continuation of work—very important and rare for a builder. He will therefore often cut costs to a minimum in the hope of being successful, resulting in a very competitive selection of tenders. A builder should be able to cut his costs knowing that, as the projects are built, his team should become more efficient so the productivity rate should be raised, also his organisation will be attuned to the work owing to its repetitive nature and, of course, any new plant purchased is guaranteed a full working life for the duration of the projects. A safeguard for the client is generally entered as a clause in the contract, stipulating that if the work is not up to standard, subsequent contracts will be withdrawn.

The package deal

This is so termed because of the all-in service it provides. In a contract of this nature, only two parties exist: the client and the builder, who will provide all the services, such as architect, quantity surveyors, consultants, etc.

Under the package deal, a client will provide brief details as to his requirements and either select a builder or offer it to open or selected methods, leaving the onus of design and costs to the builder. Generally this will concern the industrial type of building and may leave the client little choice of design, resulting in his not always getting what he hoped for. More critically, the client has no independent consultant to advise him on design or costs. On the other hand, the possibility of extra payments over the tender figure should be nil and completion should be to time as no problems should arise due to various organisations taking part. Costs should be low because of the repetitive work of the company.

Much work may come to the builder through other ways than that of tendering or the various methods already dealt with. One of the most likely sources is, of course, the 'reputation' of the builder, which has taken possibly many years to build up, and all members in the organisation have to play their part in achieving this, from management to operative.

Another method closely related to a firm's reputation is that of receiving work on the 'recommendation' of a satisfied client; this can also result in work of a continuous nature such as maintaining various properties: for example, banks, supermarkets and the like, after showing competency for work of this type. Speculative or 'spec. building', as it is often termed, is now quite common and this generally means that the builder takes a gamble (usually a very calculated one) of building houses, office blocks, etc., before having any client.

These methods of obtaining work all hold a place in the makeup of a progressive company and to create a steady flow of work all must be exploited to the full.

Performance bonds

To ensure that the employer – often public bodies – will not suffer any losses should the contractor through any reason be unable to complete the work he has tendered for, a performance bond may be taken out at the request of the employer.

This simply means obtaining a suitable independent guarantee, usually from a bank or a company, known as the *Surety*, who will, if called upon to do so, bear the cost of the unfinished work of the contractor up to the full price of the bond limit.

Once issued by the Surety this bond cannot be cancelled other than by agreement with the employer. In the event of non-performance the maximum financial benefit, which in the UK is usually 10% of the contract value, will be awarded by the Surety to the employer, giving no benefit to the contractor as it is not an insurance policy.

Generally these bonds are subject to counter indemnity which the contractor gives to the Surety, having the effect of limiting the Surety's liability in the case of the contractor's business failing.

It is therefore necessary for the contractor to be informed by the employer that such a bond is to be taken out, so that this liability and any charge for the service offered by the bank or insurance company can be added to the contract tender price.

8 Contracts

Generally a contract does not have to be in any special form, but may be written, given by word of mouth, or by conduct, gestures, etc.

For any contract to be valid it must contain various essentials:

1 There must be an offer and acceptance.
2 There must be an intention to create legal relations.
3 The contract must be either under seal or consideration: each party to the contract must receive some sort of benefit or gain from the contract. The exception to the rule is the law relating to 'sealed' documents—'signed, sealed and delivered'.
4 The parties must have capacity to contract (capacity in law, e.g. of age, of sound mind).
5 There must be genuine consent by the parties to the terms of the contract, e.g. not entered into by mistake, fraud, etc.
6 The contract must be legal and possible.

In the absence of one or more of these essentials, the contract may be void (one which the law regards as no contract at all, e.g. mistake, incapacity), voidable—one which is valid unless and until it is treated as void by one of the parties, e.g. incapacity (people under 18 cannot enter into contract), misrepresentation, fraud, duress. Briefly, if one person (or persons) makes an offer and the other accepts, a contract of agreement comes into existence.

In the construction industry, the JCT Standard Form of Contract is used if a contract document as such is desirable.

Contracts in use within the building industry

The construction industry uses a wide range of contract documents including:

1   The Joint Contracts Tribunal Standard Form of Building Contracts, known as JCT 80, in various forms – the differences being for use by local authorities or for private use, and mainly whether Bills of Quantities or approximate quantities are to be used or not.

2   There is also the Intermediate Form (IFC 80) which is designed to be used for works of a simple content, falling between the JCT Standard Form with Quantities, and the JCT Agreement for Minor Building Works, ideal for work of a straightforward nature involving only the normally recognised basic trades and skills of the industry, without complex service installations or other such specialist work. The work however must still be adequately specified and/or billed so that proper tendering can take place.

3   The Minor Building Works Agreement is similar to the JCT 80 although in a simplified form.

There is also a Standard Form, with Contractors Design, which as the title suggests is used for projects of a design and build nature, differing only in special clauses prepared to meet this type of contractural arrangement, many clauses being similar to those in the JCT 80.

4   Other forms include those for Nominated Sub-Contractors (NSC/4), which relates to the JCT 80 as to nomination procedures.

Due to the big increase in labour-only work in the construction industry, the Labour Only Subcontract Form is widely used. It is published by the Building Employer's Confederation in conjunction with the JCT, who produce the Domestic Form of Subcontract (DOM/1) for non-nominated subcontractors when the main contract is JCT 80.

Government contracts are governed by the General Conditions of Government Contracts for Building and Civil Engineering Works, GC/Works/1, and is issued for major works carried out for government departments. Form GC/Work/2 is used for minor works. The Property Services Agency of the Department of the Environment is responsible for the general content and updating of both these forms.

Management contracts are now being widely used and basically refer to the process where the main contractor provides the management of a project, performing the duties of organising the building team and the building process. For this role he would receive a fee which will include everything, even profit. Subcontractors are the main labour resource in such contracts and tender for the work (each trade independently) which should result in a low cost project. We therefore have a contract in which the main contractor provides the management skills while carrying out none of the construction work and receiving nothing of any profits made.

Why use a contract document?

Because of the complexity of building and the co-ordination and co-operation of so many separate parties, problems and misunderstandings are bound to occur; many of these (thank goodness!) are resolved by using discretion and good judgement amongst those concerned.

To lay down guide lines and standards of procedure to cover many of these problems the Joint Contracts Tribunal, consisting of:

The Greater London Council and Local Authority Association;
The Royal Institute of British Architects;
The Royal Institution of Chartered Surveyors;
The Building Employers Confederation;
The Chartered Institute of Building;
The Committee of Association of Specialist Engineering Contractors;
The Federation of Association of Specialists and Sub-contractors;
Association of Consulting Engineers;

has produced the JCT Standard Form of Contract, covering many items laid down in legal terms in a comprehensive form of contract.

Contract documents

These consist of the contract drawings, the specification and the schedule of rates. To ensure that copies are always available (at all reasonable times) they remain in the custody of the architect.

When the contract has been signed, the contractor is given by the architect without charge:

1   one copy of the Articles of Agreement—certified on behalf of the employer
2   two copies of the contract drawing, and
3   two copies of the specification

The contractor on his part must furnish the artchiect (for the employer) with a schedule of rates upon which the contractor's estimate was based, no charge being made for this document.

During the course of work when new drawings or details are required, the architect again without charge will furnish the contractor with two copies to enable the contractor to carry out and complete the work. To enable the architect or his representative on site visits to see documents, one copy of the specification and drawings must be on site. Upon completion of the project after final payment the architect may request the contractor to return all drawings, details, specifications, descriptive schedules and any other documents of similar nature that bear the architect's name.

9 Planning

Planning aims to lay down the direction in which a move is made forward, taking into account the resources that are available. Nothing can result without a plan in any form of production as activities will not be related and persons will go their own ways—the result: chaos and disorder. Any plan produced can only be a guess as to what is likely to happen, but by the use and collection of all relevant information and its critical examination, this guess becomes a much more calculated one, with a high degree of accuracy and foresight of problems and possible delays. Above all it will enable set standards to be laid down which will produce a control system. The planning used in the building industry is varied and considerable but the basic outline of types and methods have been laid down in this chapter in the order in which they are used.

Policy planning

Any organisation, to survive, must look ahead to the future, to assess trends, markets and finance; this is generally the task of top management, the principal, board of directors, or, in some cases, a committee. This activity is termed policy planning and in most cases the success of a company will depend a great deal upon its correct assessment. The policy of a company should be laid down in writing, and will cover the following topics:

a the objective of the business: what will the company do?

b financial structure, e.g. give within specific limits the total working capital of the company and what returns are expected in profit of this capital. This is important so that the danger of overtrading or taking on too many projects can be foreseen.

c time scale: Policy should be looked at, at very regular intervals, to ensure that the forecast is going ahead as planned:

e.g.: if a contractor went into a new sphere of operations, great profit would not be expected at the beginning, but if the new activity lost money and kept on losing money, at the end of the allotted time it may be better to withdraw than continue and lose.

d the overall activities of the business, e.g., to be general building contractors, to produce own joinery, to take on painting and property repairs within 10 miles of head office, etc.

e purchases: a builder, being basically an investor in plant and materials, must lay down a strict policy on the purchasing of such items as plant, formwork, equipment, with reference to hire or buy as required.

f organisation: the general set up of the business, with the approximate allocation of heads of departments; at the same time stating the personnel policy the business is to follow in such items as training, promotions, pensions schemes and the like. Great thought must be given here to keep overheads down to a minimum yet remain efficient.

Pre-tender planning

As already stated in the chapter on tendering, much of the builder's work is obtained through the process of tendering and if a builder is to stay in business, his estimates have to be competitive and a good percentage of them successful. It therefore follows that in the preparation of an estimate for work, all facts that are possible to gather should be collected and critically examined for this object to be achieved. Everyone's help should be sought in the various departments of the business to make sure that expert advice is readily available; in this way the risk of inaccurate decisions is reduced.

Even the procedure of pre-tender planning needs to be planned so that a systematic approach can be made to ensure that all information and facts are gathered on time. A typical example of this procedure is as follows:

1   The pre-tender report (Site Investigation)
2   Method statement
3   Plant schedule
4   Site organisation structure and site on-costs
5   Sub-contracts and suppliers
6   Outline programme
7   Final estimate (for Board's decision)

Pre-tender report: A document compiled to show in as comprehensive a form as possible all information regarding the area and general site conditions. An example of a report is shown and is usually carried out under the control of the planning department. The estimator may of course add questions that he deems necessary relating to a particular project, on to the standard document which is usually used and developed by the individual company. This is not a task that can be carried out in a hurry and great care is necessary to ensure that all facts are reported and that nothing is overlooked.

A. N. OTHER LIMITED
PLANNING DEPARTMENT
SITE INVESTIGATION REPORT

PROJECT

PREPARED BY   DATE

I SITE

1   GENERAL DESCRIPTION
2   LOCAL AUTHORITY
3   ACCESS

4 CROSSOVERS
5 TEMPORARY ROADS
6 DISTANCE OF SITE FROM MAIN ROAD
7 WORKING SPACE FOR SITING OFFICES, ETC.
8 TRESPASS PRECAUTIONS
9 POLICE REGULATIONS
10 CONCEALED SERVICES
11 NEAREST BENCH MARK
12 PHOTOGRAPHS

II SUB-STRATA

1 TYPES OF SOIL
2 STABILITY
3 ANTICIPATED WATER TABLE
4 SOURCE OF WATER
5 PUMPING
6 DISPOSAL OF WATER

III SERVICES   AUTHORITY   NEAREST SUPPLY

1 WATER AUTHORITY
2 ELECTRICITY
3 GAS
4 TELEPHONE
5 TV

IV LABOUR

1 AVAILABILITY
2 LABOUR

V TIPPING FACILITIES

VI LOCAL SUB-CONTRACTORS

VII OTHER SPECIAL DETAILS

Method statement

Basically this indicates how the project is to be built, what plant is to be used, and so on. Each stage of the operations is studied to find the best method of completing it, carefully weighing the various alternative methods that could be adopted. Consideration has to be given as to whether we require the cheapest or fastest method, very often not one or the same, for it may be expedient to spend a little extra at one stage to finish early so that other operations can be started. It is therefore necessary to have the operations in rough sequence before method statement is started.

A section of a method statement is shown, Fig. 8a.

Plant schedule

This is carried out on the completion of the method statement and is a detailed summary of all the plant and equipment required to build a project, giving as much relevant detail as possible as shown in Fig. 8b.

Site organisation and on cost: These are the site overheads that will have to be charged to the project. They are in fact items that cannot be priced in the course of normal productive work and will include: the site staff required for administration; temporary roads and hoardings; site huts; offices and stores; power and water; telephone etc. Site supervisory or technical staff will be calculated in relation to their administrative time. If some of their time is spent productively, this can be calculated in the normal way; for example, a craft foreman may be allowed 50% of his time for administrating his section; the other 50% would be on production. An example of site oncosts is shown in Fig. 9.

Sub-contractors and suppliers: On most projects today sub-contractors are to be found. It is of great importance in this pre-tender stage that a full list of required sub-contractors and suppliers is drawn up. It will then become necessary to

METHOD STATEMENT

CONTRACT ....   PREPARED BY ....
CONTRACT No ....
SHEET No ....   DATE ....

Activity	Method	Plant Output	Remarks
EXCAVATION Site Stripping	Bulldozer	20 m³	Soil to be moved approx 50m³
Foundation and Drain trenches	J.C.B 0·25m³ bucket load direct into 0·5m³ dumper	10 m³/hr	Cart to spoil heaps at North Side of site
CONCRETE Foundations	200NT Mixer with loading hopper, concrete discharge direct to barrows	1·2m³/hr	Only required for first 10 weeks

a

PLANT SCHEDULE

CONTRACT ....   PREPARED BY ....
CONTRACT No ....
SHEET No ....   DATE ....

No.	Description	Weeks	Availability Own	Availability Hire	Maintenance	Remarks
1	Tractor type D4 Bulldozer	1	–	✓	–	
1	J.C.B 0·25 m³ bucket.	1·5	✓	–	Daily and weekly routine	
1	200.N.T. Bedford	10	✓	–	– Do –	

b

Fig. 8

SITE ON COSTS

CONTRACT—Single stores block
CONTRACT No. 123/68   CONTRACT PERIOD—19 weeks

SITE STAFF PERSONNEL   FOREMAN   J. Green

General Foreman	19 weeks	
Site Foreman. Brickwork	10 weeks	
Site Foreman. C & J	8 weeks	
Quantity Surveyor	3 weeks	
Site Clerk	18 weeks	
General Office Cleaner	19 weeks	Recommend use of O.A.P.

SITE HUTMENT AND ACCOMMODATION

No 1.	Foremans Office	19 weeks	Phone—Fully Equipped
No 1.	Clerk of Works Office	19 weeks	Phone—Fully Equipped
No 1.	General Office	19 weeks	Phone—Fully Equipped
No 1.	Operatives Hut	19 weeks	Table & chairs
No 1.	Plant Stores	5 weeks	
No 1.	General Stores	19 weeks	Include lock up box
Latrines		19 weeks	
Compound		19 weeks	20m x 20m open mesh

NOTICE BOARDS No.2.
CONCRETE HARD STANDING (MIXING BAY)

PLANNING DEPT.
PREPARED BY: H.A.C.
DATE: 25 JULY 87

Fig. 9

send out an enquiry form (Fig. 10) not only to obtain quotations but to ascertain whether or not the sub-contractors can be available on dates required and can complete work in whatever time has been roughly allocated on the pre-tender programme. More detailed questionnaires are generally sent out to select sub-contractors if the tender is successful, for the purpose of obtaining more definite information for producing a master plan. Enquiries must be sent out to suppliers in much the same manner as sub-contractors but with more consideration as to delivery dates and prices. Care at this stage must be taken to ensure that supplies can be maintained regarding quantity and quality.

Outline programme: Often referred to as pre-tender plan. This should be produced as soon as possible, from estimated information. This is not a detailed plan but outlines the main operations and sub-contractors' work. The plan will enable all parties concerned in the pre-tender preparation to co-ordinate their activities and to assess times required for plant, use of site huts and supervisors' duration on site. Fig. 11 illustrates a pre-tender programme.

Estimate: With all the information now available, the estimator can begin to allocate prices to items in the Bill of Quantities, estimating not only the direct cost, e.g. labour and material, but also the indirect costs, e.g. site supervisory staff, general overheads, etc.

Upon the completion of the estimate, the estimator will present the persons concerned with finalising the tender, e.g. board of directors, construction manager, with a summary of the estimate with costs, enabling them to give approval or otherwise.

After the final completion and vetting of the estimate, a figure is agreed and submitted. The estimated price can now be put forward as the tender figure. This price when received by the architect is not in any way a form of contract but will be referred to as the contract figure when the successful builder has been selected.

SUB-CONTRACTORS QUESTIONNAIRE SHEET

JOB NO _____ FIRM _____ TRADE _____ DATE _____

QUESTION	ANSWER
1 How long will it take you to complete the whole of the work?	
2 What is your proposed sequence of work?	
3 What labour strength do you intend to employ on the site?	
4 How many visits will you have to make to complete the work?	
5 Do you require additional information before commencement on site?	
6 What is the minimum notice which may be given for commencement?	
7 State your storage requirements	
8 What attendances of facilities do you require?	
9 Do you agree to comply with conditions of the main contract?	
10 Do you agree to enter into the standard BEC form of sub-contract?	

Fig. 10

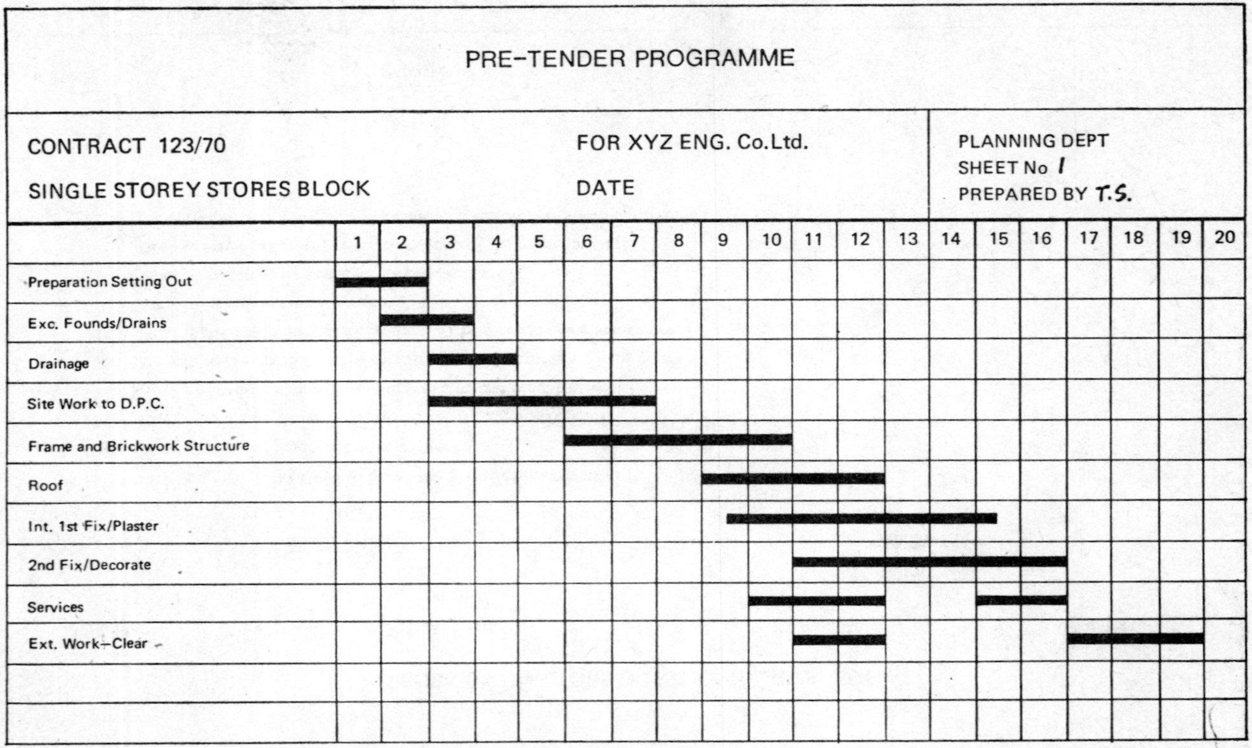

PRE-TENDER PROGRAMME

CONTRACT 123/70

SINGLE STOREY STORES BLOCK

FOR XYZ ENG. Co.Ltd.

DATE

Activity	1	2	3	4	5	6	7	8	9	10	11	12	13	14	15	16	17	18	19	20
Preparation Setting Out	■	■																		
Exc. Founds/Drains		■	■																	
Drainage			■	■																
Site Work to D.P.C.					■	■	■													
Frame and Brickwork Structure							■	■	■	■										
Roof									■	■										
Int. 1st Fix/Plaster											■	■	■	■	■					
2nd Fix/Decorate											■	■	■	■	■	■				
Services										■	■	■	■	■	■	■				
Ext. Work—Clear												■	■	■	■	■	■	■	■	

Fig. 11

Pre-contract planning

If the builder is successful in his tender and is asked to undertake the project, upon signing the contract documents he will be allowed a short period of time to make preparation and organise his resources before actual commencement of work. This period of time is called the pre-contract period and will vary in time with the size and nature of the project.

Meetings are possibly the order of the day within this period: meetings between architect, client and builder, between planning department and other departments within the organisation, and so on. The new senior site supervisor will be concerned with most of them for he is the one who will have to deal with all parties and problems on site. If he can get to grips with things before work gets under way, many of these problems may never arise.

The advantage of sound pre-tender planning will now reap rewards in so far as a sound foundation of information has been laid from which the more detailed and careful analysis of data and planning required can be ascertained.

Major items dealt with at this stage would include:

1 Site layout and general organisation
2 Labour and plant requirements finalised
3 Contract programme prepared.

The use of a contract check list so that the possibility of items being forgotten is advantageous. A possible layout of such a document showing the many aspects of procedure that must be dealt with is illustrated on pp. 71 and 72.

Contract programme

An essential detail of any project is to ensure completion within the time specified. The more complex the job, the more difficult this becomes as so many more things can go wrong. To help prevent problems arising in the form of delays and general co-ordination, a master plan is prepared to show all concerned with the project what should happen, when it should happen, and by whom it is carried out. This master

A. N. OTHER

PLANNING DEPARTMENT
CONTRACT CHECK LIST

DATE
JOB NO.

	Responsibility	Action By	Date	Termination Clearance
I INSURANCES	Company Secretary			
1 GUARANTEE BOND				
2 ALL RISKS				
3 FIRE/THIRD PARTY				
Value of Contract				
Period of Contract				
4 SPECIAL				
Demolition				
Difficult Excavation etc.				
II WATER FOR WORKS	Supervisor			
1 APPLICATION				
Value of contract				
Block Plan				
Offices Supply				
Information to plumber				
III TELEPHONE	Supervisor			
1 APPLICATION				
Loud ringing bell				
IV CROSSOVERS				
1 APPLICATION				
2 HOARDINGS				
3 GANTRIES				
V SEWER CONNECTION	Supervisor			
1 Application or quote from Local Authority				
2 What notice is required				
3 Sketch of drains run				

VI NOTICES	Supervisor
1 Commencement notices to Local Authority	
2 Factory Act Form 10	Safety Officer
3 Registration of Office Form OSRI	Company Secretary

VII SERVICES	Supervisor
1 ELECTRICAL	
a Supply to Offices	
b Building Service	
2 GAS	
Services as necessary	

VIII SIGN BOARD	Supervisor
Sign required	
Architect details	
Sub-Contractors' names	

IX SAMPLES	
Name of suppliers	Buyer
Type required	
Dates for approval	

X FIRST AID	Safety
As Construction	Officer
Regulations	

XI SITE RECORD BOX	Planner
As check list	

plan can be prepared by the use of several forms of planning techniques and will result, whatever method is adopted, in a visual diagram of project activities and other related information. The outline programme worked out in the pre-tender stage will be used as a foundation in the preparing of the more detailed master plan. The decisions and careful examination of all aspects of the work at this stage will most certainly result in the saving of money and time in the project period.

Preparation of programme: The more information available to the planner at this point, the more reliable and accurate will be his forecast and resulting programme. The use of the bills

81

of quantities to prepare a list of programme elements generally proves very satisfactory. To assess the duration of an element the itemised details produced by the quantity surveyor in the bill will be collected under the general heading. For example, under the heading of 'Brickwork' would also be collected damp-proof course, reveals to openings, placing of air bricks and lintels, etc. It will be realised that these items will have an effect on the output per hour in laying bricks and it is the planner's job to assess this output. This will occur for all project elements and takes the form of a calculation sheet, an example of which is shown, Fig. 12.

Information: To enable the planner to make a realistic assessment of the operation, he will call upon the historical information and records obtained from past projects (this is one of the main reasons for information being fed back to head office). Having determined the element time, starting and finishing date, it is necessary to communicate this information. This is generally done by setting out the master programme in the form of:

1   Bar or Gantt chart, or
2   Arrow network (Critical Path diagram).
3   Line of balance

Whichever method is used, certain characteristics of a good plan are essential to both:

1   It should be based on clearly defined objectives
2   It should be simple to understand
3   It should be flexible, so that alterations and alternatives can be made
4   It should provide standards, so that control can be maintained
5   It should provide a suitable balance of work, so that labour, once off the job, need not return
6   All resources should be used to their fullest extent, and not left standing waiting for other activities to be completed

Bar or Gantt chart

These are used widely in the industry and have proved to be a successful means of communication. A typical example of a 'bar chart' is shown in Fig. 13 showing the information that

CALCULATION SHEET

CONTRACT: .... PREPARED BY..................................

CONTRACT No: ....

SHEET No: .... DATE: ....

Activity	Quantity	Output	Prod. Hours	Lab/Plant Requirement	Duration in Days	Remarks
Conc. Founds	58 m³	1.2m³/hr	48	200 NT Mixer 6 Labs	6	
Conc. Bases	48	1.2m³/hr	40	200 NT Mixer 6 Labs	5	
Conc. Oversite	82 m³	1.2m³/hr	68	200 NT Mixer 6 Labs	8.5	
Brickwork 6" D.P.C.	170 000	70 Bk/hr	243	100 T 3 Bks 2 Labs	10	
Blue Brick D.P.C.	4350	60 Bk/hr	72	100 T 3 Bks 2 Labs	3	

Great care must be exercised when considering outputs so that realistic targets can be set. An invaluable guide is the information obtainable from records of past work of a similar nature, alternatively work study figures.

Cost must also be considered when method of work is being decided upon, to ensure that the method chosen is the most economical in the circumstances.

Fig. 12

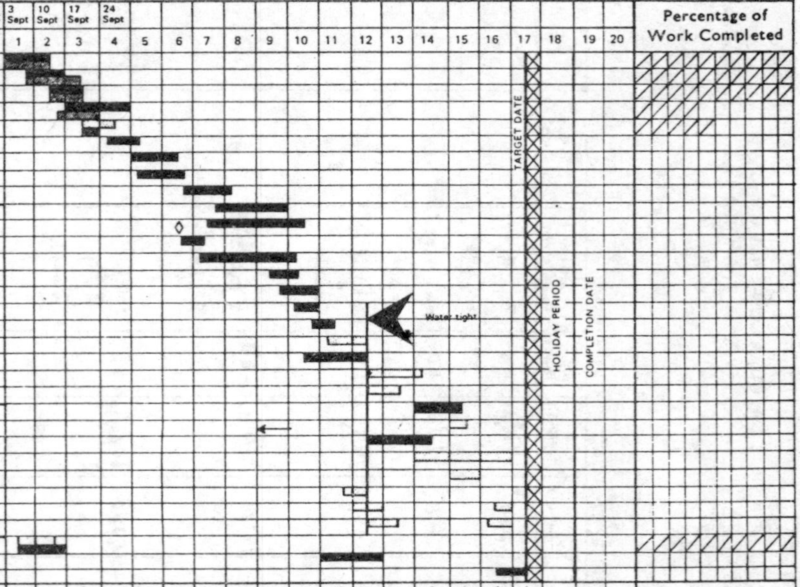

could be readily seen on such a chart. It should be realised that, as with most things, this outlay will vary from firm to firm.

Typical information that could be found on a bar chart are listed:

Contract heading

Week commencing/dates

Sequence of operations showing start and finishing dates (progress should be able to be indicated)

Labour and plant requirement

Weekly estimated labour and plant requirements

When building is watertight

Target date

Completion date

Percentage work complete section, as bar will show time spent on work, not what has been completed.

Holiday periods

Sub-Contractors

Key of items, e.g. schedules required
                       drawings required
                       sample required etc.

The difference between target completion date and completion date will vary from job to job. A general figure for projects of up to 1 year duration would be about 10% of project time; this acts as a buffer period. If a builder aims to finish on target time and does, he will gain in on-costs, overheads, etc., and will also be able re-use men, plant and equipment on other work. If he finishes on completion date, no advantage has been gained. If, however, the project takes longer than completion date, extra cost will be involved, profit lost, and also there may be a time penalty clause in the contract.

Network diagrams (critical path method)

This method is a comparatively new form of planning used in the building industry. Network analysis was first developed to ensure the successful completion of the United States 'Polaris' submarine project to time. This was a mammoth task, calling for the control and co-ordination of hundreds of main and sub-contractors. The resulting network programme proved so successful that the submarine was in service three years ahead of schedule.

The 'critical path' network analysis is only one of various types of network that can be used, but seems to suit the requirements of the builder.

Critical path planning: It must be understood that critical path method does not solve all the problems of the builder but it does help to spot and highlight some of them so that corrective measures can be taken. The calculation of element duration and resources required is basically the same as with the bar chart, using realistic and not optimistic values.

Introduction to C.P.M.: The diagram resulting from this method of planning is in the form of a 'network' or 'arrow' diagram. When the sequence or logical order of activities has been worked out, the duration times for each separate activity can be calculated and a critical path found (a path that governs project duration).

Advantages of C.P.M. over bar chart:
1 A big advantage to the programmer is that the job sequence or logic (order of things) can be completely divorced from the time element in the preparation of the plan.
2 The critical activities are clearly shown and can be altered easily if other activities become critical because of delays.
3 The non-critical activities can, with due examination, result in a more economic use of resources.
4 Non-critical activities can be delayed or performed more slowly so that resources may be used for more critical events, provided that they are not delayed so long that they in turn become critical.
5 The inter-relationship between all the activities is clearly shown by the flow of the network.

Principles of network construction: The sequence or logic diagram is set out by a series of connecting arrows, the planner questioning each activity as follows:
1 What other activity must be completed before this one can start?

2   What other activity can be done at the same time?
3   What activity cannot start until this one is completed?

Activities

These are represented by an arrow of any length. Each operation will have its own arrow. The length of the arrow has no bearing on the duration of activity, therefore an activity lasting one day could be shown longer than one lasting ten days, Fig. 14a.

Events or nodes

These are at the junction of arrows a circle is drawn to signify the completion of the activity. An event has no duration, Fig. 14a. Inside this circle is placed the identity number of the activity. This enables operations to be identified easily by their start and finish numbers instead of lengthy titles. These are termed i–j numbers.

Sequential activities

These are activities that can only proceed in correct order one after another, Fig. 14b.

Parallel activities

Activities which can be carried out independently of others at the same time, Fig. 14c.

Dummy activities

These are represented on a diagram in the form of a broken arrow line, Fig. 14d. Its use is to show the inter-relationship between activities, giving a logical sequence. In itself it is not an activity and therefore has no time duration.

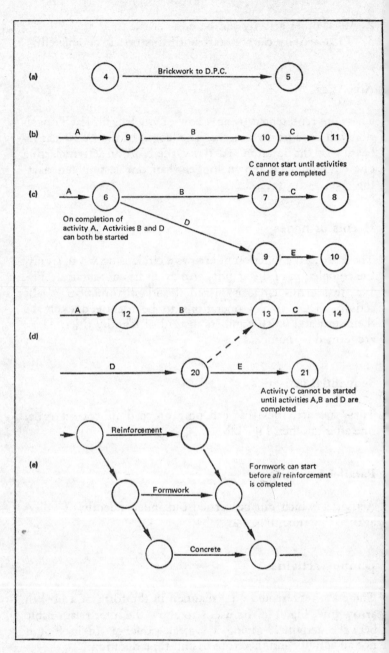

Fig. 14

Restraining activities

All activities do not have to wait for preceding activities to be finished before they can be started, as for example in the construction of concrete *in situ* columns, when it is not necessary to put all the reinforcements in every column before formwork can be placed, neither is it likely that concreting will not start before all the columns are surrounded by formwork. The method of showing that work can be started before previous work is only partly complete is similar to Fig. 14e.

Preparing an arrow diagram

1   *Logic :* The first step to consider referring to the principles already stated is the logic, or, in other words, the procedure or order in which a project is to be tackled. This is arrived at by discussion with all concerned until a suitable order of activities is agreed, remembering that duration times at the moment are not taken into account.

An example of a planning sequence is illustrated, Fig. 15a. The shape of the diagram or actual length of lines may be drawn in a variety of ways but as long as the logic remains the same, this will have no effect on finished diagram.

2   *Time :* The next stage after the logic sequence has been agreed is to add in the time elements to suit the project (e.g. hours, days, weeks etc.), Fig. 15b. This requires the accurate determination of activity time as in the case of any programme.

3   *Analysis Sheet :* The next operation is to produce an analysis of event times as illustrated in Fig. 16 to see along which path of activities the critical line must be drawn.

The earliest time an event can occur is known as the 'Earliest Starting Time' (EST) and is the first calculation to be worked out; this is done by taking into account all preceding activities on the logic diagram and adding them together. For example: from Fig. 17 it can be seen that event 3–4 cannot be started until events 0–1, 1–2, 1–3, 2–3 have been completed, remembering always that the longest duration times or longest path on the logic diagram is the earliest starting time for preceding

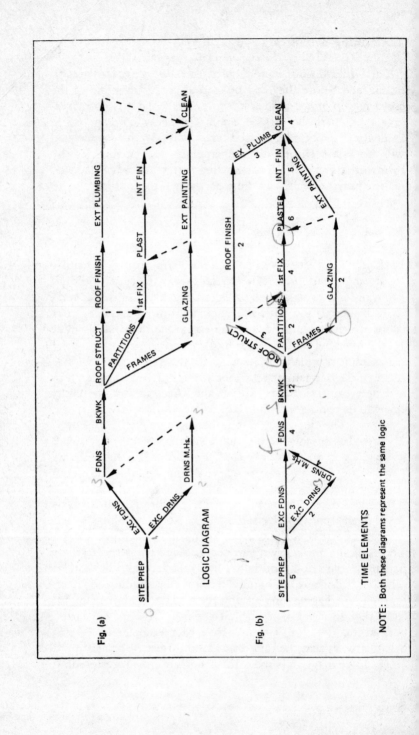

Fig. (a)

LOGIC DIAGRAM

Fig. (b)

TIME ELEMENTS

NOTE: Both these diagrams represent the same logic

Events i. j	Description	Dur.	Earliest Start	Earliest Finish	Latest Start	Latest Finish	Critical Path	Total Float
	ACTIVITY SHEET							
	JOB		EVENT TIMES					
0-1	Site Preparation	5	0	5	0	5	✳	0
1-2	Exc. Drain Trchs	2	5	7	5	7	✳	0
1-3	Exc. Founds	3	5	8	8	11		3
2-3	Drains & M. Holes	4	7	11	7	11	✳	0
3-4	Founds	4	11	15	11	15	✳	0
4-5	Brickwork	12	15	27	15	27	✳	0
5-6	Roof Structure	3	27	30	27	30	✳	0
5-7	Partitions	2	27	29	28	30		1
5-8	Frames	3	27	30	29	32		2
6-7	Dummy	—	—	—	—	—		—
6-11	Roof Finish	2	30	32	40	42		10
7-10	First Fix	4	30	34	30	34	✳	0
8-9	Glazing	2	30	32	32	34		2
9-10	Dummy	—	—	—	—	—		—
9-13	External Painting	3	32	35	42	45	⌐	10
10-12	Plaster	6	34	40	34	40	✳	0
11-13	Ex. Plumb	3	32	35	42	45		10
12-13	Int. Finish	5	40	45	40	45	✳	0
13-14	Clean	4	45	49	45	49	✳	0

If the project is a large one, with possibly hundreds of activities, the figures may be fed into a computer for calculation; the critical activities and float would be indicated.

Fig. 16

activities. The reason for this is simply that the shortest time activities can be done in the longest time but the longest path of activities cannot be done in the shortest time.

e.g.   Event 0–1 = 0 days EST
         Event 1–2 = 5 days EST
         Event 1–3 = 5 days EST
         Event 2–3 = 5 + 2 = 7 days EST
         Event 3–4 = 5 + 2 + 4 = 11 days EST.

Activity 3–4, therefore, cannot start until day 11, for this is the shortest time that all the preceding activities can be completed. Path 0–1, 1–2, 2–3, not Path 0–1, 1–3 as this only takes 8 days and 11 days' work is to be done before activity 3–4 can be started.

The latest starting time: This is simply the latest time an activity can start to be completed on time. This is found by the identical process used in finding earliest starting time, only now work backwards through the diagram from the completion date. From the analysis sheet, Fig. 16, the critical activities can now be shown. An activity is critical if the earliest and latest starting times are the same, e.g. events 0–1, 1–2, 2–3, etc. If a difference in time does occur between the two times, as in the case of event 1–3, this time is called the float; in this instance, event 1–3 = LST 8 − EST 5 gives 3 days. There are various forms of float that can be calculated from the analysis sheet. These require a deeper understanding than it is intended to illustrate in this book. It is sufficient to know that the float will enable:

1   Resources to be used to the full
2   Rearrangement of labour to complete more critical activities if production falls below set rate

Fig. 17 shows a completed arrow diagram, indicating how other relevant information can be shown. As site supervisory staff may not be familiar with CPM a bar chart may be produced from arrow diagram to show activities in a similar form, Fig. 18. Another advantage of the bar chart at this stage is that float can be shown more clearly than on network.

Allocation of resources: The allocation of resources to each activity is tackled in much the same way as with the formulation

METHOD OF ILLUSTRATING INFORMATION ON NETWORK DIAGRAM

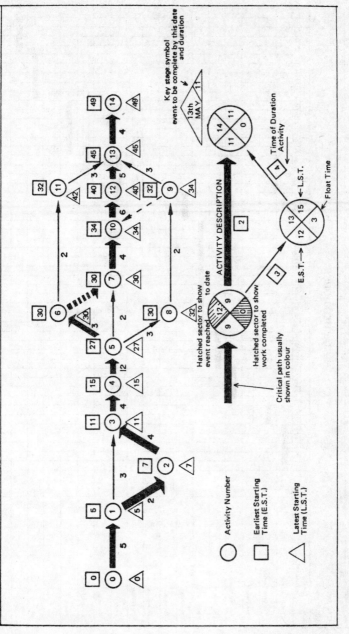

Fig. 17

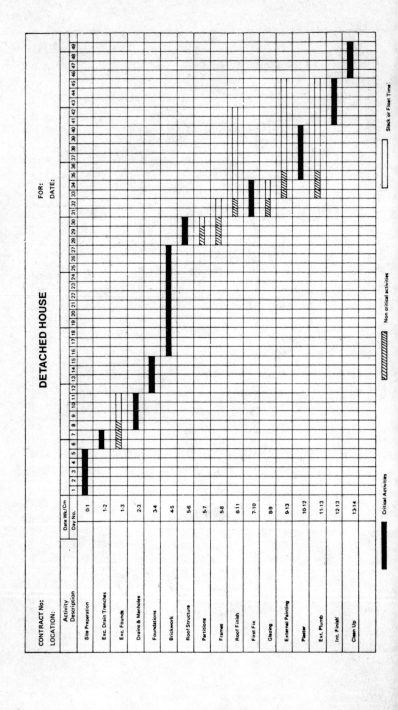

DETACHED HOUSE

CONTRACT No:
LOCATION:
FOR:
DATE:

Activity Description	Date Wk./Cm Day No.
Site Preparation	0-1
Exc. Drain Trenches	1-2
Exc. Founds	1-3
Drains & Manholes	2-3
Foundations	3-4
Brickwork	4-5
Roof Structure	5-6
Partitions	5-7
Frames	5-8
Roof Finish	6-11
First Fix	7-10
Glazing	8-9
External Painting	9-13
Plaster	10-12
Ext. Plumb	11-13
Int. Finish	12-13
Clean Up	13-14

Critical Activities   Non critical activities   Slack or Float Time of a bar chart. The main aim is to achieve a balance of plant and labour to suit activities without big variations or unnecessary movement from site to site, and is considered when logic is decided.

Line of balance

This method of planning, although long used in factory production line situations, has only been adopted to any great extent in the building industry since the early 1970s, and has since proved very effective when applied to repetitive work, especially housing and flats with similar floors.

The aim of line of balance as with other types of planning is to eliminate non-productive time, thus substantially reducing the period taken to build units. It is achieved by ensuring that identical operations are carried out repeatedly on successive units by, if possible, the same operatives, thus producing a flow of work similar to that achieved on assembly lines where repetitive operations are carried out continually. The only real difference between factory and site is that, whilst in factory production the work comes to the operatives, on site in construction work the operative goes to the work.

At commencement of a line of balance programme it follows a simple logic network as indicated in Fig. 19 which illustrates the basic elements only and a calculated time allocation for each section, making up a period in which a single unit is to be completed.

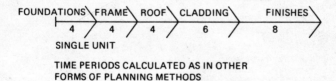

FOUNDATIONS \ FRAME \ ROOF \ CLADDING \ FINISHES \
4 / 4 / 4 / 6 / 8
SINGLE UNIT

TIME PERIODS CALCULATED AS IN OTHER
FORMS OF PLANNING METHODS

Fig. 19

It is reasonably obvious that all sections will not take the same time to complete, neither can we assume as in any other type of production that problems in the form of stoppages or delays will not occur. Also, whilst on a factory production flow line 'buffer' stocks can be built up so that if one section falls behind others can still carry on using stored stocks, this is not possible in construction work. It is, however, still necessary and so a 'buffer' is *created* using *time* as opposed to material products. Fig. 20 illustrates how the time buffers are added into the logic diagram.

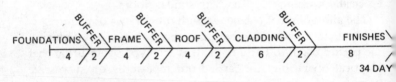

THE TIME PERIOD OF EACH 'BUFFER' WOULD BE CALCULATED IN PRACTICE FROM EXPERIENCE, CONSIDERING THE DIFFICULTY AND POSSIBLE RISKS INVOLVED WITHIN EACH SECTION.

Fig. 20

The next stage, that of assessing resources, can now be carried out. Fig. 21 shows a typical labour force pattern. The programme can now be prepared.

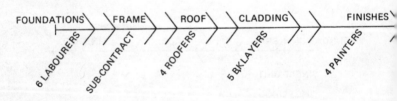

GANG SIZE ALLOCATION FOR THE COMPLETION TO TIME OF ONE UNIT

Fig. 21

96

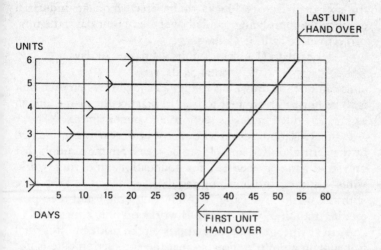

Fig. 22

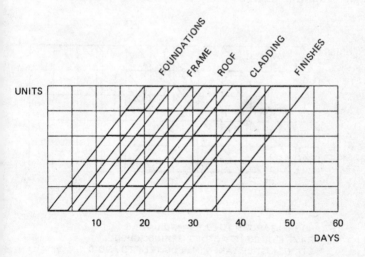

Fig. 23

If a simple example is taken using the logic diagram Fig. 20 for six small factory blocks to be erected on an industrial estate with a unit being handed over every four days, the total project period will be 45 days.

$$34 \text{ days (1 unit)} + 5 \text{ units} \times 4 \text{ days} = 20 \text{ days}$$
$$\text{total } 54 \text{ days}$$

This can now be drawn up as Fig. 22 showing when each unit is to be finished from first to last. In practice the plan is drawn as Fig. 23 which indicates sections and buffer times.

Estimated labour resources can now be drawn in, represented by the vertical dotted lines. It can be seen from the example that only one gang is required for foundations and roof work, whilst cladding and finishing requires two teams to keep pace with the work flow, gang A completing unit 1 and then proceeding to unit 3, whilst gang B works on unit 2 then 4.

As in any programme, alterations can be made but it is vital to maintain a uniform flow so that the completion date is not exceeded. Progress can be checked by using a date line, filling in as indicated in Fig. 24, which will show directly which units are on time, ahead of or behind schedule, making it possible to make adjustments to correct the situation immediately.

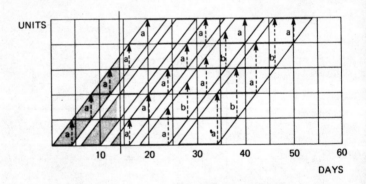

PROGRESS SHOWN TO DAY 14 INDICATING
FRAME (SUB-CONTRACTORS) BEHIND SCHEDULE
ACTION MUST BE TAKEN IMMEDIATELY TO AVOID
DELAYS

Fig. 24

Short term planning

During the course of a project, problems, delays, alterations, shortage of labour, bad weather, etc., may seriously affect the running and completion of the job. This will be shown in the loss of progress on the master plan. If we are to keep up with this plan it is necessary to consider each operation in fine detail so that we can see and evaluate any problems; this is achieved by taking a 'Forward View' of work to be carried out over a shorter period of time, generally 3–5 weeks. This period will alter depending on type of project and various other circumstances. The main aim of this short term planning is to keep the master plan up to date and 'alive' and not just a highly decorative chart on a wall that serves no purpose because things are not going as forecast. It is too late at the end of a job to find you are six weeks over time.

Preparation of short term programme: As the duration to be covered is now reduced to a few weeks, much greater detail and accuracy is possible. Each month the site agent or general foreman, often with the help of the contract manager, will prepare this programme; the operations that are behind, together with operations taking place, are carefully examined to assess how the best results can be obtained, and what alterations will have to be made to keep up with the master plan.

The assessment of labour, plant and materials, plus the careful co-ordination of sub-contractors, is a vital part of short term planning if work is to be carried out economically, so that the careful balancing and phasing in sequence, the various operations and labour are essential. The reason for preparing programmes every 4 weeks for a 5 week period is to give an overlap at the end of each period because of difficulties that would be met if programmes were end on; it also provides a buffer period.

A typical short term programme is illustrated, Fig. 25.

Weekly planning

Meetings are often held, generally on the Friday of each week, on site; the object being to ensure that progress is being

Photo copy ततिS—

STAGE PROGRAMME

CONTRACT:
CONTRACT No.

PERIOD
WEEK No. 1 TO 3

Quantities			Labour				Plant			Operations		Week 1 – 3rd Sept.						Week 2 – 10th Sept.						Week 3 – 17th Sept.					
Amount	Unit	Out-Put Hrs	Total Hrs	Gang Size	Gang Hrs	Days	Type	Hrs	Days	No.	Description	M	T	W	Th	F	S	M	T	W	Th	F	S	M	T	W	Th	F	S
			97	3	32	4				1	Temp. Fencing	3 LAB																	
			96	3	32	4					Administration and Site Hutments	3 LAB																	
			60	5	12	1·5					Open Mesh Compound							2 LAB											
			32	2	16	2					Concrete Hard Standing					5 LAB													
365	m³	10		1			JCB	36	4·5	2	Exc. Foundn.					1 LAB		1 LAB											
200	m³	10		1			JCB	20	2·5		Exc. Drains.																		
54	m²	1·2		5			TNT	44	5·5		Conc. Foundn.							2 LAB				1 LAB		4 LAB					
			48	2	24	3				3	Exc. Man Hole											5 LAB							
5050	8/hr	208	95	2	43	5·5				4	Build Man Hole													2 BKS + 1 LAB			1 LAB		
			176	4	44	5·5				30	Drains incl. Back Fill													2 LAB			6 LAB		
				1							Water / Elect. Service							LAB											
												6						6						6	2	1			

NOTES
Gen. Labourer
Bricklayer
Brick Labourer

Sub-Contract

maintained, that targets set the previous week have been accomplished and new targets set. This meeting is usually run by the senior site supervisor. All trades foremen plus representatives of required sub-contractors should attend.

To ensure its success a high degree of co-operation and team spirit is most esssential so that suitable and workable targets can be built up for the coming week. These plans are usually in the form of a simple bar chart, showing the daily work load of each gang, Fig. 26. Although other types are used that incorporate either:

a an outline of the block plan of building, useful when dealing with multi-storey construction as work load can be readily shown where it is happening

b similar to block plan method but the outline is now shown in isometric, one floor over the other, and is useful to show not only what is to be done by the gang but also the flow of work using a series of flow arrows.

Control

In any situation where progress is to be made, some form of plan must be laid down for others to follow. At the end of an allotted time, be it day, week or year, the checking of the estimated targets against the actual performance achieved will indicate the measure of control.

Accurate planning is therefore vital if effective control is to operate to ensure that all units can be continually checked, to ensure that any deviations can be suitably resolved, whilst still maintaining control over original objectives. In the construction industry, with all its possible variations, control is often difficult to ensure. In broad outline, the various sections requiring control are:

Labour	Quality
Materials	Cost
Plant	Sub-contractors
Progress	

Labour control

Maintaining sufficient labour strength whilst ensuring that the various gangs are so balanced that the work-flow proceeds

WEEKLY PLANNING SHEET/PROGRESS CHART

CONTRACT: A.B.C. LTP Factory Extension.
LOCATION: WILLS LANE

DATE: 13 Feb 1979
Supervisor: J. Black

PREVIOUS WEEKS PROGRAMME	COMP. YES/NO	REASONS IF NO.	ACTIVITY LOCATION	23 16-2-70 MON	17-2-70 TUES	18-2-70 WED	19-2-70 THURS	20-2-70 FRI	21-2-70 SAT	WEEK NUMBER DATE DAY REMARKS
Plastr. Partitions Main Block	No	Partition not complete	Complete Plastering Part. Mens Wash Room	////	////	////	////		////	
Floor to shop 3	Yes	—	Floor to shop 4 - 5	////	////	////	////	////	////	
—	—	—	Wall tiling - ladies Washroom	////	////				////	
		NOTE	Alternative method of indicating work plan	Plaster Part.		Mens Wash Room				

smoothly is of utmost importance on any construction work. It is also necessary to ensure that the tasks set for operatives or gangs are carried out within the planned requirements and that the recording of operatives' hours and progress is recorded.

Sub-contractors

It is the builder's responsibility to ensure that the progress of the sub-contractors is maintained to the requirements of the master plan. The key words to achieve success here must be co-operation and co-ordination. The sub-contractor should be kept fully aware of his obligations whilst given every assistance in the performance of his tasks.

Progress control

The indication of how much progress has been made relies upon the degree of detail required in relation to the programme of which the work is being carried out, e.g. weekly for master programme, daily for weekly programme. Progress is usually shown in a physical manner by using a colour to block out or underline the estimated duration of task. Another method is to show by means of a percentage column how work is progressing.

Conclusion

Sound forecasting is necessary to produce realistic programmes that can be used by site staff to control projects. Other documents aids are also used to achieve this end and will vary with size, complexity, duration of project, but may consist of:

    Plant schedules
    Sub-contractors' schedules
    Requirement schedule for drawing of details
    Labour requirements, charts, etc.

Whatever form of document is used it must be simple to understand, realistic in its forecast and, above all, up to date.

10 Budgets and costs

Just as site production is programmed so that progress can be checked, i.e. estimated time against actual time, so must the cost of actual work be checked against estimated (tender) price. Expenditure must be programmed for it is of little use being ahead of a production programme if the project is losing money.

To observe overall costs, budget graphs are drawn up. These may take the form of a collective graph including labour, plant, materials and overheads, or, alternatively, separate graphs can be prepared; this latter method is possibly the better. With a collective graph, it cannot be seen which section is running high or low to its estimated expenditure.

Preparation of a budget graph

Having the master programme now prepared, labour requirements estimated, materials and plant assessed, a weekly or monthly *estimated* cost can be calculated. To this must be added such items as overheads, on-costs, insurance, etc. These items are simply calculated in total and divided by project duration.

$$\text{Example:} \quad \frac{\text{overheads—insurance—on-costs, etc.}}{\text{project duration (weeks or months)}}$$

A typical example of a budget graph is shown, Fig. 27.

The *actual* expenditure each week/month can be calculated and recorded against the estimated costs.

Unit costs

Just as a master programme is ineffective unless detail in the form of short term planning is used, so budgetary control is only constructive if item costs can be checked as to whether or not profit has been made.

The estimator will have calculated a cost of each unit of work to be carried out, e.g. brickwork £32.00 m^2 or excavator £3·85 m^3.

Actual costs

It is now necessary to calculate the actual costs; these are obtained by using site records similar to the daily allocation sheet as shown in Fig. 28. These items are collected on a weekly summary and finally taken to a weekly cost control sheet, Fig. 29. It is usual only to cost control labour and plant as this is the area with the greatest variability.

Various systems of costing are in operation but all aim to produce the same end result:
1   keep a close check on overall expenditure
2   control unit costs
3   provide suitable feed-back for future use.

Cost control

The record kept on costs will be of great value to the estimator when determining future prices so that good documentation and feed-back is essential. The main object of costing is to obtain costs in the appropriate units of measurement and should be related to all items of expenditure on a project: labour, including bonuses, etc., materials, plant, transport,

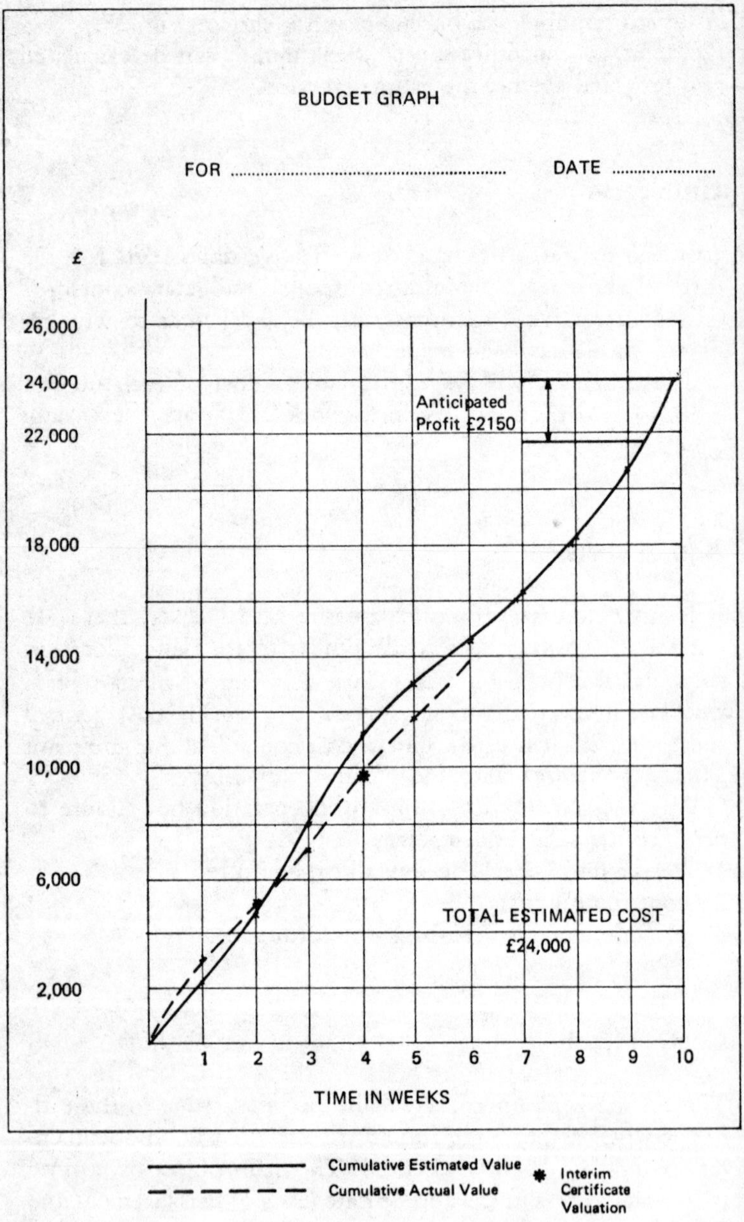

BUDGET GRAPH
FOR .... DATE ....
£
26,000
24,000
22,000
18,000
14,000
10,000
6,000
2,000
Anticipated
Profit £2150
TOTAL ESTIMATED COST
£24,000
1
2
3
4
5
6
7
8
9
10
TIME IN WEEKS
Cumulative Estimated Value
Cumulative Actual Value
Interim
Certificate
Valuation

Fig. 27

A.N. OTHER Ltd.   Date ....

DAILY LABOUR ALLOCATION

CONTRACT ....
CONTRACT NO ....
SUPERVISOR ....

Op. No.	Name	Trade	LOCATION AND DESCRIPTION OF WORK	Hours On Each Operation								
Total Man Hours												

Fig. 28

OPERATIONAL HOURS ARE COLLECTED FROM THE DAILY ALLOCATION SHEET SO THAT THE TOTAL HOURS FOR EACH OPERATION IS PRESENTED.

THE COST CONTROL SHEET SHOWS ALL THE INFORMATION COLLECTED FROM THE VARIOUS SITE RECORDS ENABLING UNIT COSTS TO BE CALCULATED.

WEEKLY SUMMARY SHEET

CONTRACT
CONTRACT NO

WEEK NO
WEEK ENDING

OPERATION No.
OPERATION

TYPE OF PLANT USED

Day	M	T	W	Th	F	S	Total
Lab. Hrs							
Trade Hrs							
Plant Hrs							

OPERATION No.
OPERATION

TYPE OF PLANT USED

Day	M	T	W	Th	F	S	Total
Lab. Hrs							
Trade Hrs							

A.N.OTHER

COST CONTROL   SITE   CONTRACT No   W/E

No.	Description of Work	Unit	Rate	Quantities		Value		Cost This Week			Cost Cumulative			Total	Unit	Remarks
				Week	Cumte	Week	Cumte	Lab	Plant	Total	Lab	Plant	Total	Cost	Cost	

sub-contracts payments, site on costs. Costing systems vary from firm to firm, generally in relation to the size of the project. Careful and accurate compiling of cost figures is the overriding process.

11 Site layout and organisation

To ensure that a site is laid out in the most effective and efficient manner, much thought and consideration must be given to the task before the work actually starts. The period after tender acceptance and actual starting date, called the 'pre-construction period', is the time to carry out this exercise.

The arrangement of site huts, storage sheds and compounds and the placing of loose materials, mixing bays and so on, will more than likely be the first consideration on the site and the last removed on project completion. This orderly layout shows new operatives that they are employed by an organisation with planning care and consideration in mind. To the general public, the impression created is one of efficiency, which is a good advertisement. Through the period of a project, a good tidy site will often reflect the efficiency of the site supervisor, reduce waste and purposeless movement of plant and materials. Therefore the visualising of the project through all its stages must be envisaged to fulfil this purpose.

Adjoining property

Many sites abut or are surrounded by other buildings, and it is important that good relations should be maintained with the owners of these properties right from the start. No harm is done by putting people in the picture as to what has to be done and the stressing that as little inconvenience as possible will be the aim. This preliminary information may help to reduce friction if problems in regard to dust, noise, etc., occur during the course of the project; the surrounding owners will more likely be prepared to meet the builder half way in the case of difficulties, instead of taking stronger, more direct action. It is often practicable also to insure against mishaps and suggested damage to existing property by taking photographs (getting owner to sign as correct) and placing tell-tales over any faults to avoid legal battles later on if owner accuses builder of damage.

Access to site and traffic routes

Every contractor should remember that although particular circumstances of each site or job will determine the extent to which recommendations can be applied, the builder should provide:
1 such drainage required to keep site reasonably free from standing water
2 so that access to parts under construction provide roads or paths
This in itself should be sufficient to ensure that the site is suitable to work on. Many projects, especially in towns and built-up areas, often have to make new access to sites by crossing over existing footways, and whether the access is of a permanent or temporary nature, the permission of the local authorities must be obtained.

Another hazard that often occurs in these areas, especially if the site is of a restricted nature, is the problem of unloading materials or loading up spoil and rubbish for removal, as this may cause interference with the flow of traffic. Often police help in this matter is advisable and necessary. Various ways may be put forward to get over the problem such as:
   a the construction of a 'pull in'—temporary layby
   b the introduction of a one-way system past the site, controlled by temporary traffic lights or a manned 'stop–go' board

c the delivery of materials at 'off peak' periods

Whichever the method adopted, adequate warning should always be given to other road users by the display of notices, danger signs, warning lights or other suitable communications. If sites are fully enclosed, the use of sliding doors for access is much more advantageous than standard opening ones.

Contractor board

This should be clearly displayed, remembering again that this is an advertising aid, so that traffic and visitors have no problems in finding the site. In country districts the use of 'finger boards' is very helpful in directing people to the site as street names are not always easy to locate.

Site huts

The W.R.A. and the Health and Safety at Work etc. Act lay down minimum requirements to cover operatives' welfare facilities on site, but much will depend on size and nature of the site as regards other provisions.

Generally sites will consist of hutments that can be divided into three groups:

a Administrative offices
b Operatives' huts
c Storage and general

a The administrative offices will, as stated, vary depending upon the size of the project, from a single hut used by site supervisor and possibly visiting clerk of works, to a complex of offices to house site supervisors, engineers, surveyors, planners, time clerks, resident clerk of works, and many others. The materials used to construct these offices will also vary but in most cases are now designed of sectional timber or other suitable material that can easily be handled, transported and erected. All types should be of sound construction, watertight, with adequate lighting and ventilation and suitable internal finishes and fixtures to impress visitors to the site and provide suitable surrounding for meetings. Offices should all be linked, from a simple sliding panel to a sophisticated intercom system, for ready communication. On congested sites, two-storey offices are available, and in some cases the office may have to be situated on a gantry above ground level—generally to allow access for transport to site if narrow frontage.

b Operatives' huts; the W.R.A. minimum recommendation for these are: shelter from inclement weather; accommodation for clothing; accommodation and provision for meals.

Most builders realise the benefits that the extra costs involved in applying the recommendations of the W.R.A., and in many cases improving on them as improved conditions produce: good relations between operatives and management; better morale, team spirit, which in turn will result in higher productivity; reduced labour turnover which will give a more knowledgeable team with reduction of retraining; higher productivity due to operatives working better if dry clothes can be put on at the start of each day, and if hot meals can be obtained to replenish energy lost. Shelters and canteen should be kept clean. The use of an old-age pensioner here usually produces satisfactory results, besides releasing a more active operative for other work—in many cases, unfortunately, an apprentice. An adequate number of tables and chairs should be provided so that rests and meal breaks can be taken in comfort. The standard of internal furniture should be considered—the use of easily cleaned tables with tops of formica or similar are to be recommended, with the use of metal framed chairs with suitable seating. This will certainly help create a feeling that the management is considering the men. The position of these huts is again a matter for careful consideration and whenever possible they should be placed in proximity to the administrative block (foreman's office) so that an eye can be kept open for operatives taking prolonged breaks or other unscheduled visits to huts during the day.

Sanitary conveniences: are covered in great detail by the W.R.A., including provisions for the washing of hands. Unfortunately this is still an aspect of site welfare that is often of a poor standard, but with the aid of mobile ablutions better facilities should result. Drinking water must also be provided.

c Storage and general: Depending upon conditions, other general hutments may be necessary, for example if site and contract is large, employing many operatives, there must be a properly constructed and maintained ambulance room. Also on a large project, it may be necessary to have a manned stores, a time-keeper's hut and even a material checker's hut at the entrance to the site, for checking and directing material supply. To ensure satisfactory layout of the site as mentioned in the opening paragraphs, two simple methods can be employed to ensure best results; both entail the use of a plan of the site showing the outline of building with drain and service runs (to scale if possible).

1 Cover with sheet of plastic and draw an arrangement, simply rubbing out wrongly placed huts, and so on until the best solution is achieved, or:

2 Prepare simple plywood or hardboard cutouts to scale, and place these around the plan until satisfied of the layout, Fig. 47.

By using these methods, the wrong positioning of necessary allocated area is reduced to a minimum and the moving of plant, materials and huts because of unforeseen service trenches is eliminated.

Check lists

At most stages of any operation, in any size or type of organisation, regular checks should be made to ensure a smooth flowing sequence of activities and events. This is the reason why it is good policy to provide a check list of items required at the commencement of site operations, even as early on as site layout; this will insure against items being forgotten or not on hand when required. Two such check lists are illustrated, Fig. 30, a check list of equipment necessary for commencing a project, whilst Fig. 31 shows a list of necessary documentation required on site, which can be delivered in a box with the list enclosed for checking off the items.

A.N.OTHER LIMITED

PLANNING DEPARTMENT

CHECK LIST AND REQUISITION FOR NEW CONTRACT

JOB NO.

DATE

Qty.	Details	Load No	Qty	Details	Load No
	Barrows			Level, dumpy	
	Battens			Materials inwards boards	
	Bonjng Rods			Notice Board	
	Bolt Croppers			Nails	
	Brooms, bass			Pegs	
	Brooms, soft			Picks	
	Buckets			Ranging rods	
	Building Square			Rubber boots	
	Calor Gas ring			Setting out lines	
	Calor Gas boiler			Shovels	
	Calor Gas cylinders			Sign board	
	Cement Shed			Spots	
	Chairs, folding			Staging	
	Crow bar			Scaffold tubes	
	Electric fire			Scaffold fittings	
	Forks			Sleepers	
	Forms			Spirit levels	
	Hammers, lump			Stools	
	Hammers, sledge			Table	
	Hoardings			Tapes, steel	
	Hose pipe			Tarpaulins	
	Huts (size)			Tool box	
	Huts (canteen)			Wash bowl	
	Ladders			Wash stand	
	Lamps, danger				
	Levelling board				

Load No. 1 required by
Load No. 2 required by

Fig. 30

A.N. OTHER LTD

PLANNING DEPARTMENT
SITE RECORDS BOX

Date.......................   Job No.................

CHECK LIST

I   **FACTORY ACT FORMS AND REGISTERS**
   1961 1580 General Provisions
   1974 Health and Safety at Work etc. Act
   1961 1581 Lifting Operations
   1966   94 Working Places
   1966   95 Health and Welfare
   F.91 Part 1 Scaffolds
   F.91 Part 2 Lifting Appliances
   F.36 General Register—Young Persons etc.
   FB1. 510A Accident Book
   F.2202 Shared Welfare Arrangements
   F.988 Woodworking Machinery Regulations
   F.05R.9B Offices, Shops and Railway premises
   F.1 & 3 Factories Acts
   SHW.279 Safety Hints—Woodworking machinery
   F.954 Electricity Special Regulations
   R.11 Permissible hours of work

II **CONTRACT INSTRUCTIONS**
   5 Weekly Planning Sheets
   Weekly Planning Sheets
   Internal & External plant hire rates

III **FORMS AND RECORD SHEETS**
   Site diary
   Weekly reports
   Site instruction records book
   Daywork book
   Daily return book
   Dimension book
   Drivers' log sheets
   Accident report forms
   Fresh starters record book
   Loose plant transfer pad
   Materials transfer pad
   Plant transfer pad
   Petty cash pad
   Requisition pad
   Sub receipts book
   Wages book
   Wages sheets

IV **STATIONERY**
   Carbon paper—A4, A5
   A4, A5 paper
   Folders
   Envelopes, addressed, A4, A5
   Duplicate book
   Internal memo pad
   Scrap pads
   Ball point pens
   Pencils
   Paper clips
   Pins
   Thermometer

Fig. 31

12 Materials

Supply procedure

No construction project can proceed without a satisfactory supply of raw materials; so that besides the careful phasing and planning of material required by the builder, it is to his advantage to foster a good relationship with his suppliers, many of whom will have been selected due to their fulfilment of orders to the standard required and meeting of delivery times over a number of years.

Besides the builder's own suppliers, the architect may specify that a certain supplier must be used and these are termed 'nominated suppliers'. Whatever the type of supplies to be used, the information passed to them and received from them is the same, and in all but the smallest firms this information and documentation will pass through the buyer. It is also the responsibility of the buyer to ensure that the architect receives any samples from the suppliers in the very early stages of contract procedure to satisfy him of the relative merits of the material. It may, for example, not be possible to obtain the specified material in time in conjunction with the building programme, but by obtaining samples of similar products the architect may decide on a new form of construction or design to prevent holdups.

In most cases, a buyer will send enquiries to two or three

suppliers or in some cases direct to manufacturers for such items as sand, gravel, bricks, cement, etc., regarding prices, delivery dates and such, for use at the estimating stage of the project.

This will enable the estimator to use the figures obtained in the preparation of the tender figure. If a builder is then successful in his bid for the work, he will place his orders with the suppliers of his choice by comparing the various quotations received in relation with the following:

1 Has the supplier been used before?
2 If so, was he reliable?
3 Is the material suitable for the project?
4 Can the material be maintained in quantity?
5 Delivery dates
6 Price including: a Trade discount
                  b Cash discount
                  c Bulk discounts
7 Do materials come up to British Standards or other tests that may be required?
8 Any special considerations

Upon these points proving satisfactory, a definite order will be placed and, upon acceptance, a contract is formed. A typical order form is shown, Fig. 32. To enable quick checks to be made on delivery or non-delivery of materials, a card index system is generally adopted; this enables the buying department to keep a constant eye on supplies.

These cards contain all necessary information and are best filed in date order for ease of reference. Even when bulk deliveries have been phased according to programme, by taking the extra time and trouble to write out a card for each delivery the chance of losing sight of the delivery date is very much reduced, as regular checks will be made on the file. As materials are received, the information will be recorded on the card.

Schedules

An aid often used in the ordering of materials is the schedule; this is usually produced by a quantity surveyor or by a material

A.N. OTHER LTD.
    BUILDING CONTRACTORS
    1762 BLOCK Rd.
    REDHILL, WAR.

Tel. 041 6996 341

	No. **8999**
	PURCHASE CODE
	/ /
<u>ORDER</u>	
	Date
	ALL INVOICE —
	statements and correspondence
	relating to this order should be
	addressed to our Head Office
By accepting this Order you agree to the terms and conditions printed on back.	QUOTING THE ABOVE NUMBER AND DATE.

Please carry out work/supply in accordance with your quotation numbered ....and dated .... as follows:	PRICES

Deliver to:	Commence.................	For and on behalf of **A.N. OTHER LTD.**
	Complete.................	
	see conditions 2.6.13	————————— Director

It is most important, when making out an order, to specify and state clearly exactly what is required. On the back of most order forms, the terms and conditions of the order are stated, and range from such items as insurance against loss, and liability, to bankruptcy and failure to complete order.

Fig. 32

scheduler, by systematic analysis of the Bill of Quantities and contract drawings; a typical example of a schedule is shown in Fig. 33.

Ordering procedure

The process of ordering is shown, Fig. 34, remembering that documents relating to ordering should always be in triplicate or even more depending on the system:
Top copy: to supplier.
Second copy: either to yard storeman or to site.
Third copy: retained for filing by the purchasing department. A fourth copy may be sent to the accounts department for checking invoice.

Documentation in supply of materials

Advice Note: this is sent by works or supplier to site, stating date, method of transport and description of goods despatched. This will enable site supervisory staff to make adequate preparation for unloading and storage. Fig. 35.

Delivery Note: Fig. 36. This is the document that must be signed by the foreman when he has unloaded the goods stated on the note and is satisfied with their condition. Careful check must be made to ensure that goods are not only there but are also in good order and not damaged. Often goods arrive in packing cases or crates and it is not possible to examine the contents; if this happens, upon signing the delivery note the foreman should also write across it 'Not examined'. Two delivery notes will be supplied by the transport driver: one for himself as a check to his employer that he has made the delivery satisfactorily and one for site reference.

If goods are missing or damaged, this should be clearly shown on the delivery note and an appropriate letter sent as soon as possible to the supplier. If goods are returned for any reason with the delivery lorry, the driver should be asked to sign correction on the delivery note.

DOOR SCHEDULE

SITE *PLUMB ROAD*   DWELLING TYPE *A/37*

| DOOR REF. | LOCATION | SIZE | FACE FINISH | NO. REQUIRED | PRE-MORTICE | | ADDITIONAL INFORMATION |
					LOCK OR LATCH	LETTER PLATE	
17/3 Ax	Bathroom	1980mm x 812mm	Oak ply both sides	6	Mortice latch	—	75 mm Barrel bolt
24/7T	Bedroom	– Do –	– Do –	24	– Do –	—	

SCHEDULE OF FINISHES

SITE...............

FLOOR No......

ROOM	FLOOR	WALLS	CEILING	SKIRTING	DOOR	ADDITIONAL INFORMATION

DRAIN SCHEDULE

SITE...............

| LOCATION | | AV. DEPTH | CON-CRETE | DIAM. PIPE | BENDS/ Junctions | CHANNEL/ GULLIES |
FROM	TO.					

MANHOLE SCHEDULE

SITE No...............

M.H. No.	INTER SIZE	COVER	INVERT LEVEL	COVER LEVEL	BASE	SUNDRIES

Fig. 33

Excellent means of communication.
Information in an easy to understand form.

MATERIAL SUPPLY PROCEDURE

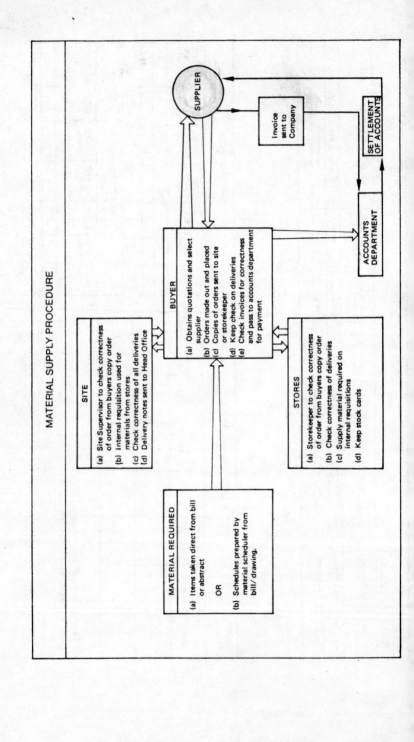

SUPPLIER

Invoice sent to Company

SETTLEMENT OF ACCOUNTS

ACCOUNTS DEPARTMENT

SITE

(a) Site Supervisor to check correctness of order from buyers copy order
(b) Internal requisition used for materials from stores
(c) Check correctness of all deliveries
(d) Delivery notes sent to Head Office

BUYER

(a) Obtains quotations and select supplier
(b) Orders made out and placed
Copies of orders sent to site or storekeeper
(c) Keep check on deliveries
(d) Check invoices for correctness and pass to accounts department for payment

STORES

(a) Storekeeper to check correctness of order from buyers copy order
(b) Check correctness of deliveries
(c) Supply material required on internal requisitions
(d) Keep stock cards

MATERIAL REQUIRED

(a) Items taken direct from bill or abstract

OR

(b) Schedules prepared by material scheduler from bill / drawing.

H&S BAT LTD.

No. 448905

TEL. 555 680 1136/9

148 HIGH St.
WATERLEY
SUSSEX

ADVICE NOTE

┌─ INVOICE TO

┌─ DELIVERED TO
(Customers address unless otherwise stated)

Despatched by:	*Lorry*	Consignment No.		Date
Customers order	*764/69A*	Date		

QUANTITY	DESPATCHED	RETURNS

IMPORTANT   If the above are not received within TEN DAYS from the date of this ADVICE NOTE kindly advise us, otherwise we cannot accept responsibility.

Careful note should be made regarding conditions of sale that often appear on the back of this document regarding such items as: delays, price increases, defective materials, loss etc.

Fig. 35

<u>DELIVERY NOTE</u>   NO 5069

Received From

J. K. BLACK LTD

PHONE 786 908 78/9

BUILDERS MERCHANTS

STANFORD GREEN, NORTHINGTON.

MESSRS .... YOUR ORDER NO ....

.... DATE ....

Please receive the undermentioned goods.

REMARKS: RECEIVED BY:

Remarks similar to those given below may be found on delivery notes:

The goods specified above are delivered on terms of business set out in our quotations, price-list, and/or hire conditions and are accepted accordingly.

Complaints of shortages or damage cannot be recognised unless notified in writing to Company Head Office within *3 days* of delivery, stating (a) Ticket No. (b) Date.

Fig. 36

Invoice: These are very similar to delivery notes, inasmuch as they state the same information, the only difference being that the price of the goods is now clearly shown for payment. This document is sent to the head office where it is checked against the delivery note which has been sent from the site. If they agree, payment will be made. Fig. 37.

Requisition: This is a document used by the site to obtain sundry items from the central stores on a day-to-day basis. Fig. 38.

The important consideration here is that transport is not wasted in bringing a few small items, when with a little thought most items can be brought in one good load.

Materials record book: This is a complete record of all materials received on site and must be filled in before delivery notes are sent to head office. Entries should be made on the day of delivery. Fig. 39.

Materials delivery board: Often used on site to show clearly when deliveries are due, general bulk phased deliveries, quite often simply a ruled blackboard on which is chalked amount and date of delivery.

Material transfer: Fig. 40. This document is a company document for internal use when materials are moved from one site to another for any reason. Its correct use ensures that charges for materials can be made to the receiving project and deducted from the site supplying the goods. In the accounts, the site giving will be credited, the site receiving will be debited.

Waste materials

A big problem on most building sites is the large amount of material that, due to varying circumstances, becomes classed as waste. Basically this is a problem of the site supervisory staff to control and wherever possible eliminate; it requires a supervisor to be constantly on the lookout for loss

Little success will be achieved if the operatives do not play

COLD & SNOW

INVOICE

257 BLEAK STREET
NORTHINGTON
YORKSHIRE

TEL. 356 789 5500/4

INVOICE NO. 443119
PLEASE QUOTE THIS REFERENCE
VAT REGISTRATION NO 179 6754 02
A/C NO.
INVOICE DATE

ACCOUNT TO

DELIVERED TO

Your Ref.No.	Quantity	Description	Price	Value	
For important conditions see reverse side of this INVOICE			GOODS		
			VAT		
			INVOICE TOTAL		

An important document which must be checked, before payment, for correctness regarding delivery from delivery-note and prices from quotation.

Fig. 37

A.N. OTHER LTD

INTERNAL REQUISITION

To ....

From .... Contract No. ....

Date .... № **66780**

Please supply or Order for the above Site/Department, the following which are required by .... (enter date)

Quantity	Enter full Description of Materials

FOR OFFICE USE ONLY

Date Received	Action by		
		Order placed	
		Order No.	
		Goods delivered	

Each site would be issued with a book of Internal Requisitions. Two copies would be made out on each request for material—top copy to store, duplicate retained.

Fig. 38

| A.N. OTHER Ltd. | | | | Date .... | | |

DAILY RETURN   B 7705

Delivered To	Ticket No.	Date	Material	Office Use Only		
					£	p

This book is filled in daily from information taken from the delivery notes. The prices and amounts are filled in against each item at Head Office. A running total cost of materials for a project is easily and quickly obtained.

Fig. 39

A.N.OTHER LTD
BUILDING CONTRACTORS

MATERIAL TRANSFERS

To ....

From ....

Department or
Contract No. .... Co. ....

Department or
Contract No. .... Co......

Date ....

No. 68003

Quantity	Enter full Description of Materials	For Office Use Only	

FOR OFFICE USE ONLY

Date	Co	Main Code	Sub-Code	DR. Contract	CR. Contract			DR. COSTED
Priced By								
								CR. COSTED
Extended By								
		Total Per Transfer Note						

Similar to Internal Requisition but passed to another site and not to stores, duplicate copy kept by site.

Fig. 40

their part and so it should constantly be impressed upon them the importance and value of all materials. A number of firms adopt a visual approach to this by placing upon a large sign board on site various items of material, such as a brick, tie iron, roofing tile or length of timber, etc., stating alongside each its cost, and so drawing attention to the fact that even very small items do cost money. In most cases, it is not the waste or loss of valuable items that gives rise to concern, simply because these items are generally very carefully stored, checked and issued when required. It is the everyday items of material which is the biggest concern. Generally, because each small separate item costs very little, sight is lost of the fact that if a large number is wasted, the cost can be very high. Take for example the brick—in itself not of much significance; but add together the number that are broken in handling, lost through bad storage, thrown under lorries that get stuck, buried under aggregate piles; besides those lost through the cutting of halves—and this adds up to a high percentage of the total and thus a loss of money. This applies to many such materials. How, then, can waste be reduced?

a Ensure that materials are delivered as required so that site storage time is cut to a minimum. This requires careful phasing of deliveries between site and supplier

b Ensure that materials delivered are those specified for that particular job

c The issuing of just the right amount of material with only a reasonable allowance for wastage to workmen.

d Ensure that workmen are not producing excessive amounts of 'offcuts'

e Allocate and prepare storage areas. This can be done by marking on the site plan the exact layout of all material storage areas; giving each a code number or letter, and then marking this area out on site with pegs or similar, showing the equivalent code. Drivers delivering material can then be told by reference to the plan where to place their load, Fig. 41. Make sure lorries can get to the area with as little interruption to normal work as possible and that sufficient space is allocated so that other materials or work is not damaged in the process of unloading. Always try to ensure that materials are re-handled as little as possible and that

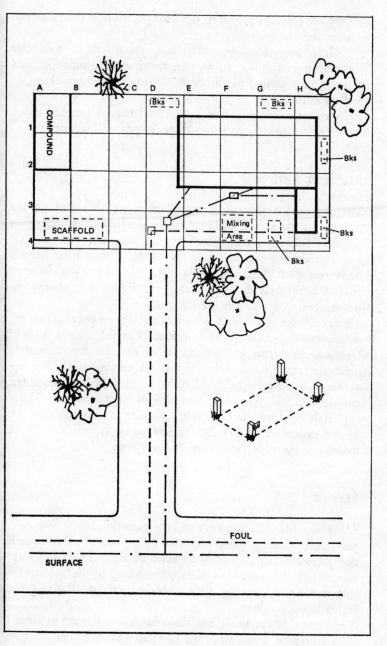

Fig. 41

they are placed as close to the final position in the structure as room allows

f Make sure that when materials are stored, they do not deteriorate. Fig. 42 shows a material storage schedule that could be supplied to all site supervisors to ensure that this does not happen

g Collect waste, e.g. half bricks, and use to prevent more cutting—a few by each bricklayer's spot will suffice.

Material control

The prime function here is to ensure that materials are ordered in good time, and that a very close watch is kept upon planned delivery dates. Schedules are required here so that quick reference can be made as to when and from whom deliveries are required. Material in short supply or late delivery should be chased up immediately, with possible alternative materials or suppliers being sought.

Control on site must be exercised in respect of waste, deterioration, pilfering and misuse. Careful checks should be made to ensure correctness of orders and that materials delivered can be properly stored and unnecessary handling avoided. Standard quality of materials should be maintained through checking against samples or specifications. Many materials used on site can be tested quite simply without the use of expensive laboratory facilities such as would only be found on the very large complex projects.

Security

Materials: Many items of building material are stolen each year by outside persons and this reflects in general overall site security. Also much material and small hand tools are lost to the builder each year through pilfering by the operatives; this is easier to control than outside theft by adopting the following procedures:

a Items that are small, e.g. door furniture, should be issued by storemen, foreman or the like and a record kept

MATERIAL	METHOD OF STORAGE							STACKING	REMARKS
	Open Site	Hard Standing or Divided Bins	Lock Up Ground or Compound	In Partially Completed Building	In Weather Proof Lock Up Shed	Delivered As Required For Use	Open Site with Protection From Rain		
CEMENT					X	X		Off Ground 5 Bags High Max.	Use in Rotation First in First Out
AGGREGATES		X				X		Prevent Deterioration and Separation	Phase Deliveries
CONCRETE BLOCKS							X	Near to Point of Use	Keep dry to Prevent Shrinkage
CARCASSING TIMBER							X	Raise of Ground stack in sizes Support Evenly	Seal Ends
WINDOW AND DOOR FRAMES			X				X	Stack in Sizes off Ground	Prevent Damage
JOINERY			X	X				On Bearers in Bundles	Ditto Keep Dry
DRAINAGE GOODS	X		X Specials			X		Tie in Ends of stacks	Valuable Liable to Breakage
PLUMBING FITTING				X	X	X All Metals		Leave in Packing Cases	Ditto
JOINERY FITTING					X			In Boxes	Keep in Sets
GLASS				X	X	X		Upright on Bearers In Sizes & Types	Keep Dry
BRICKS	X		X Specials			X		Near to Point of Use Tie Ends	Check Phased Deliveries

Fig. 42

b Accurate stock control must be maintained with regular checks

c Compounds and storage sheds should be kept locked after issue of materials

d Cars, wherever possible, should be allocated space away from the construction so that 'loot' cannot quickly be hidden in car boots

e Site supervisors must set an example by practising what is preached

Site Security can cause many problems and it becomes very difficult to lay down hard and fast rules and precautions that can be followed because of the considerable difference in sites, type of building and the firm's efficiency in these matters.

The problem is also not just one of keeping the professional criminal out but also the vandal, old and young alike, and, of course, the onlookers and curious visitors who trespass upon the site and often cause much damage due to ignorance. It is clear, therefore, that security has to guard against two separate problems:

a Theft, especially of high value materials such as copper and lead, that not only result in loss of money but may also cause delay in construction

b Damage due to unauthorised persons being on site, whether wilful or accidental

One of the most common forms of defence used on site is that of constructing a hoarding around its perimeter; this generally is sufficient to stop the curious sight-seer, but for the criminal they present little deterrent as they are usually easy to scale and, of course, once the unwelcome visitor is inside, he can work undetected as he cannot be seen through the protecting hoarding. It would seem, therefore, that some form of opening hoarding is better, such as a chain link fence, and with the addition of a few strands of barbed wire at the top reasonable protection should result, for although easy access may be obtained by cutting a way in, this is noticeable, and a movement on a locked-up site is easily spotted. Openings in either type of hoarding should be adequately locked and secured to ensure that criminals cannot drive in transport.

Inside the site, all movable items and small pieces of plant, equipment and materials should be locked away and, of course, with careful ordering, stock will only need to be kept to a minimum, which in itself is a deterrent to a would-be thief as he may consider the site not worth breaking into as rewards are small. A practice that is also growing is that of maintaining floodlighting on the site during the hours of darkness; this again helps to deter the criminal as he can easily be spotted.

The use of night watchmen is open to criticism due to the fact that most of those engaged in this work are old and would prove of little value in the apprehension of a criminal; but they are useful inasmuch as that they may disturb him, and should then have means of raising the alarm. A decrease in the use of nightwatchmen has taken place with the introduction of guard dogs similar to those on military installations, although care has now to be taken when dogs are used as conditions must comply with the Guard Dog Act of 1975.

Keys often result in a breakdown of security due to loss or inadequate control of who has them or access to where they are kept. A close check should be kept on keys and only persons requiring them should have them allocated.

As already stated, the problems of site security are many and varied but it must be remembered that often the police will be able to offer invaluable advice, especially if the site is situated in a heavy crime density area. It is also good practice to ask the police to keep an eye on the site, and also either leave with them a set of keys or a local address where keys can be obtained in case they require access through theft or fire.

Children: An increasing cause for concern is that of accidents to children on building sites.

The Health and Safety Executive estimate that about twenty children each year are killed on sites, many only five years of age and younger.

The best precaution to prevent this situation is to keep children out of sites by better site perimeter fencing, but regardless of this all reasonably practical steps should be taken as laid down in the H. & S.W. etc. Act 1974 so that not only employees but the public (including children) are not exposed to risk, even if trespassing.

13 Plant and equipment

Plant and equipment have, over the past few years, been introduced into the construction industry in an ever-increasing volume; this has, to some extent, been the reason for the increase in output whilst the labour force has fallen slightly. Besides which, it has also helped to a very large extent in the reduction of heavy manual work.

The range of plant to be found in the industry is wide and varied, as can be seen from the following list of items:

Pile-driving plant	Demolition plant
Tower cranes	Hoists
Concrete mixers	Dumpers
Excavators	Bulldozers
Pumps	Compressors
Scaffold	Power hand tools

These are just a few of the more general items a builder would use in the course of normal construction work.

To hire or buy

Because of the structure of the industry with its many small units, there has developed the Plant Hire Organisation, whose role has been to provide a service for the 'little man' who requires and wants to use items of plant but, due to circumstances, cannot afford to buy, and the larger concerns who at times require specialised plant for limited periods. These hire firms have proved to be most successful and are recognised in the same manner as any other form of sub-contractor. The big advantage in using a plant hire firm is that as specialists and experts in a particular field, a good range of up-to-date, efficient working plant should be obtainable at very short notice.

Why, then, should a builder buy if such a good service is at hand? The main reason is, of course, costs. If a builder owns his own plant, that is constantly employed, the cost will be less than that of a plant hire firm. The choice of buying or hiring is generally one of policy. A few of the more general considerations to make before purchasing plant are:

1 Will there be sufficient work for the item, not only now but over its working life?
2 What will the machine do?
3 What is not only the initial outlay, but also the costs incurred in repairs, replacements and maintenance?
4 Will transport be required to carry items from site to site?
5 Will skilled operators have to be employed or trained?
6 What is the working life of the item?
7 Power source (electrical, petrol, diesel).

Use of plant

Whether plant on site is hired or bought, certain considerations are applicable to both:

1 Ensure that the plant is clean, is in good working condition and regularly maintained. It is always good practice to give responsibility of each machine to the operator, allowing time each day for daily routine checks so that it can be kept in good order.
2 Machines are designed to do certain tasks; it is therefore necessary to get the best work out of the machine. It should only be used for the job for which it is designed.
3 Make sure that machines are safe and that there is no danger in their use to site operatives or general public.

4   Ensure that statutory records are kept.
5   Maintain records of machines for future reference and feed back. Fig. 43 illustrates plant records.

Maintenance of mechanical plant

One of the main disadvantages in the use of mechanical plant is if it breaks down it generally causes considerable hold-ups and delays: for example, a tower crane serving an industrialised building site, if this fails the site is likely to come to a complete stop. It is necessary to ensure that failure does not occur; therefore regular maintenance is essential. The large firms having plant departments carry this out in the same way as plant hire firms, usually having a set routine and mechanics to meet all demands. The smaller builder, if he owns plant, will either contract the work out or consider the possibility of appointing a fitter who, being provided with a small van, could carry out all but the major overhauls.

Whatever system is used, careful records should be maintained and maintenance carried out to makers' instructions. Careful planning is necessary here to ensure that work will not suffer due to plant being out of action for regular maintenance or overhaul.

Plant cost rates

The cost to the builder of using an item of plant on site can be arrived at from the following:

Initial cost	Items will vary with type of
Scrap value	plant—much information can
Estimated life period	be obtained from past records
Depreciation	and feed-back
Licence and insurance	
Repairs and maintenance	
General overheads	
Profit	

Example:

Cost of plant	£40 000·00
Less scrap value	5 000·00

PLANT RECORD CARD

TYPE Mobile Crane   SIZE No. 164/341   RATE £1·89 wks/day/hr

Date	Contract / Contract No.	Date of Charge	Wks Days Hrs	Charge		Date	Contract / Contract No.	Date of Charge	Wks Days Hrs	Charge	
11·3·70	HV2 L70 / 124 King Street / H₂ 123/70	14·3·70	10 hrs	18	90						

PLANT TRANSFER SHEET

Transferred From				Transferred To		
Date	Contract / Contract No.	Item		Date	Contract / Contract No.	Item

PLANT OPERATING SHEET

TYPE   CONTRACT   CONTRACT No.   WEEK ENDING

Op. No	Operation	Hours Worked						Total	Fuel	Remarks
		Mon	Tue	Wed	Thur	Fri	Sat			
Breakdown Time										
Idle Time										
TOTALS										

Fig. 43

Depreciation 20%
Interest on capital 5% (not always taken)
Repairs and renewals 33⅓%

Initial cost	£ 40 000·00
Less scrap value	2 000·00
	38 000·00
Annual depreciation 20% per annum	7 600·00
Interest on capital:	
1st yr. £38 000 @ 5%	1 900·00
2nd yr. £36 100 @ 5%	1 805·00
3rd yr. £34 295 @ 5%	1 715·00
4th yr. £32 580 @ 5%	1 629·00
5th yr. £30 951 @ 5%	1 548·00
	8 597·00
Plus original cost (total depreciation)	35 000·00
	43 597·00
Repairs and renewals, 33⅓%	14 532·00
	5) 58 129·00
Cost per year	11 626·00
Plus tax and insurance (say)	1 500·00
	13 126·00

Cost per hour

Average hours worked		
Weeks in year		52
less Holiday periods	3	
Maintenance periods	2	
Transport to site	1	
Idle time/Breakdowns	8	
	—	
		14
Total weeks working		38

Working hours per week 44
Working hours per year: $44 \times 38 = 1672$ hours
Hourly cost per year: $1672)13\,126 = £7·85$

Operating costs

Machine operator, *9 hours @ £4·50	=	£40·50
Fuel per day (say)		£7·60
Oil (say)		£2·00
Grease and rags (say)		£1·00
Therefore cost per 8 hour day	=	£51·00
Therefore cost per hour	=	£6·50
Therefore total cost per hour		
(£7·85 + £6·50)	=	£14·35
Add 10% overheads	=	£1·44
		£15·79
Add 10% profit		£1·58
Therefore final cost per hour	=	£17·35

*1 hour allowed for maintenance. Costs for fuel, oil, grease, etc., would be obtainable from past records.

Output

It is necessary when selecting plant to ensure that its output is sufficient to carry out the work in the required time (programmed targets). A typical example of how this can be calculated is illustrated.

Example: concrete to be placed in foundations in 6 days. Amount 154 m³. Actual production time *4/5 × 8 × 6 = 38·4 hours

*1/5 day allowed for non-productive time, e.g. cleaning plant, etc.

Required output $\dfrac{154}{38\cdot4} = 4$ m³/hour

Assume 300/NT (Approx. 035 m³)

Mixing cycle:	loading	1 min
	mixing time	3 min
	discharge	1 min
		5 min

Therefore 12 batches produced per hour × 0·35 m³ = 4·2 m³
Output per 6 working days = 4·2 × 38·4 = 161·28 m³
Therefore this mixer will produce the required output.

Matching plant

Often during the course of a project, various types of plant are used in conjunction with one another, as for example, lorries with an excavator, concrete mixer and wheelbarrows. Because of the high cost of plant it is most necessary to keep it working at maximum efficiency; to do this, careful matching of plant is essential.

Example of plant matching

Transport units must be such that output plant is not delayed by having to wait for one empty one, i.e.:

$$\frac{\text{Output per hour}}{\text{Transport unit carrying capacity per hour}} + 1 \text{ extra transport}$$

unit to ensure full productivity. No waiting at output.

Carrying capacity of transport unit:

Effective load per trip × number of trips per hour

Time of trip made up of: a Loading time
                          b Unloading time
                          c Travelling time

Travelling time includes   (i)   loaded outward haul
                           (ii) empty return haul
                           (iii) manœuvring in position both ends

Example: How many 0·50m³ dumpers are required to maintain a 0·50m³ N.T. mixer at maximum efficiency if concrete is to be placed at 200 m distance?

Mixer output 6 m³ per hour

Therefore $\dfrac{6 \text{ m}^3}{0·5 \text{ m}^3}$ = 12 loads per hour.

Vehicle loading time $\dfrac{60 \text{ mins}}{12 \text{ loads}}$ = 5 min

Vehicle unloading time (say) 0·50 min

Travelling time 200 m at average speed
of 15 Kmph + 0·20 min each end for manœuvring, say

$$\frac{1·50 \text{ min}}{7·00 \text{ min}}$$

Therefore number of trips per hour $= \dfrac{60}{7} = 9$ trips

Therefore carrying capacity of each vehicle $= 0 \cdot 50\text{m}^3 \times 9$ trips
$= 4 \cdot 50\text{m}^3$ per hour

Therefore number of dumpers required $= \dfrac{6}{4 \cdot 5} + 1 = 3$ Dumpers

It must, of course, always be remembered that plant efficiency is not only dependent upon the plant itself but also on the efficiency of the operator and the general conditions prevailing at any given time.

Whether plant is the builder's own property or hired, three main points must be considered:

1 Ensure that plant arrives on time. This is important so that other operations are not held up.
2 When plant has finished, it should be taken off site. In this way the site will not be charged for plant that is lying idle; it also ensures full utilisation of plant.
3 Plant progress should be recorded to ensure that planned output is being achieved. Records of plant breakdown are also necessary, especially in the case of plant hire.

Plant requirements are usually controlled by the process of preparing a plant schedule from the pre-contract method statement co-ordinating this with the site progress as shown by the master plan.

Plant Hire

There are approximately 1400 plant hire firms employing 40000 operatives and staff in this country concerned with a £600 million a year turnover.

Plant hire has become such a big part of the Construction Industry that from 1 February 1979 a new Working Rule Agreement came into effect, produced by the Contractor's Plant Association, the TGWU Construction and Crafts section, UCATT and GMWU.

This new very much enlarged agreement covers every aspect of employment for all types of plant hire operatives including general plant operators, fitters and crane drivers.

14 Quality

Quality control

An essential to any industrial organisation's standing and reputation is the standard and quality of the goods, services or work they carry out and building is no exception.

Building structures do however, create their own peculiar problems as to quality. Very few industries are affected by so many variables, for not only are there problems of labour workmanship standards, but a wide range of materials, products and components, weather, design and costs all play a part. There is also the on-site assembly to produce the finished product. Far more difficult situations arise from on-site production than factory made items, due to the difficulty of quality control supervision.

When looking at these variables it is clear that costs must be one of the most demanding when considering quality. One only has to consider the relationship between a Rolls Royce and a Mini to see how this reflects the measures of comparison, especially when we take the Rolls as being perfection, for this is what all industrial concern strive for, although seldom achieve.

As with most finished products – be it Rolls or Mini, high class office development, or a small semi-detached house, if the client's expectations to quality are met the quality has to be assumed satisfactory. Often this may simply be on the visual appearance of the finished structure but, of course, in construction many of the aspects of quality will not show themselves for some considerable time. For example, when maintenance costs have been studied, which closely reflect the durability and reliability of the structure, which in turn is reflected in the workmanship and quality standards of the finished structure.

Design

This is the aspect that in the first instance relates costs to the fabric, layout, size and shape and appearance of the structure.

The designer at this stage is always governed by the current statutory regulations, such as the Building Regulations and the Health and Safety at Work Act, as examples, when considering these restraints and those of construction time related to overall costs the designer must ensure that his presentation is not unrealistic, otherwise there is little doubt the quality of the finished product will suffer.

Details of the structure under consideration must be clear and follow good building practice, which in turn should result in a long term defect-free structure. Put as a simple equation:

$$\left. \begin{array}{l} \text{Purpose and use} \\ \text{Constructability} \\ \text{General appearance} \end{array} \right\} = \left\{ \begin{array}{l} \text{Cost limits} \\ \text{Project duration time} \\ \text{Quality control} \end{array} \right.$$

Purpose and use

Consideration will include structures, ranging from low cost housing to prestige 'one off' developments and the designer will consider the needs of the client. For example, a factory being built within very limited cost budgets may have to sacrifice high quality appearance to ensure internal service – such as insulation, heating or air conditioning – vital so the client's use of the structure is fully implemented.

The designer will also need to ensure that long term and flexibility of use, levels of servicing and maintenance, all fall within the client's brief, so that the consideration of quality which relates to all these factors is the first priority.

Constructability

In other words the assembly of all the various manufactured items and materials, with all their differing characteristics, into a complete structure. This requires the designer to know and understand the tests, standards, and performances of units to ensure that the parts are suitable for the whole.

General appearance

This may in some respects be restricted and governed by local conditions ensuring that the new structure blends with the existing environment.

The designer may have a choice, as mentioned before, to create a traditional structure or create a status building. The latter, however, often produces a more difficult process with new methods of construction and materials being used. These more complex structures may create problems for the work force making the required standard of quality difficulty to achieve, and thus require generally more time and a higher level of trained supervision.

Cost limits

The designer's dream and the contractor's workmanship in producing the finished structure should link very closely to the standards recognised as being required by the client. The designer should from the beginning indicate clear guide lines to be followed throughout the progress so that appropriate quality standards are achieved.

Low budget structures do not necessarily mean poor quality and shoddy work, but considering design, the standard of material and finish applied, it is clear a relationship does exist between them, although this is often only in the visual context of the finished structure.

Construction duration time

Whatever the project it will be covered by the client's overall budget, closely linked with design. This in turn is directly related to the duration programme of work. The higher the quality standards required often the longer the working time that has to be allowed. Whether the item is site or factory produced, one only has to consider the Rolls and Mini comparison again.

On-site weather will also influence the duration time of a project, and as it is so variable little can be done about it other than careful planning to make the best use of the periods when better weather is expected. Often contractual clauses may have to be turned to in order to resolve problems in this area.

Unfortunately, on many occasions little thought is given to the problems of weather which can cause serious difficulties in relation to quality control. Consider the effects of rain and frost on brickwork and concrete.

Achieving quality standards

Whatever the standard set it should be measureable either by testing, measuring, or simply by checking visual appearance. These standards are usually to be found in the drawing and detailing, the specification or the bill of quantities, together with the related British Standards, Code of Practice, etc.

One of the problems with these methods of communication is that generally only office staff have access to them as terms of reference, and are little use to operatives who never see them. Yet it is quite clear that if the on-site work force is not committed and motivated to produce good quality work, it will not be achieved without their use. It is, therefore, vital to overcome this shortfall by illustrating standards, either by samples or other forms of visual display. Above all the operatives must be made aware of their role by clearly defining the site quality standards. Quality control on site is largely dependent on leadership and it is vital that the site management lines of authority are clearly delineated and fully understood, so that responsibility can be taken for quality achievement.

This demands appropriate levels of site management resources, properly trained, who can exercise authority to achieve results. This reduces the problems when quality deficiencies *are* identified: someone has the responsibility and authority to take the necessary remedial action without long delays and discussion. This is becoming a far more difficult field of operation now that more and more sub-contractors are being used with the main contractor providing only the management team.

147

Inspection

On-site inspection is in the hands of the supervisor at craft level to the supervising architect. On most important projects a treble framework exists, to try to ensure that quality standards are achieved.

From the outset of construction the local authorities, or similar outside agencies such as the water, electric and gas boards, all lay down clearly defined control standards. The contractor's own supervisory staff together with the third party, i.e. the architect or his representative (often the clerk of works) will do the same. There should be a close working relationship between these parties on quality control although they all have differing interests. Clearly, all parties should have the same documentation, and more importantly they should all understand their meaning, especially in relation to quality.

This may be a problem, however, as in practice the standards often reflect the views or impression of the individuals concerned, making it vital that all communication makes interpretation as straightforward as possible.

Considerable emphasis must be placed on the quality and experience of the 'inspectors' to ensure that they understand the agreed quality standards of the designers and the client. Poorly trained and unmotivated supervisors, regardless of rank, are possibly the most significant feature of poorly executed work, reflected in the resulting low quality standard of the project.

Architect/designer

It is the role of both to be fully aware of new techniques and materials that are available, so that the best advantages can be given to the client within current statutory restraints. They should also be aware of any limitation or possible short-coming of products.

In designing a new structure they need to understand the whole process of construction and should be able and willing to relate this directly to site work, especially in respect to quality standards. Often architects have been accused of not really understanding what materials can and cannot be expected to do or achieve.

Clerk of Works

We have already discussed the framework of his duties and responsibilities in Chapter 2, but it is his *informal* assistance that relates to his role in quality control. These people are generally former trade or general foremen and are fully aware of not only the obligations of the contractor but also of the difficulties and restraints. Often, if encouraged, the Clerk of Works will be able to offer sound advice without overstepping his duties and causing contractual hazards through his lack of recognised authority.

Site supervision

This often falls to a site agent or general foreman, as it is their responsibility to define the designer's quality standards to the work force, so that they may put them in to practice.

With the use of sub-contractors they often have little or no knowledge of the work force being used, the problem of site supervision is increased, especially if his duties are more specifically progress chasing than overseeing quality control. There is also often the need to call on the service of sub-agents so that when the supervisor's knowledge to relevant standards is limited assistance is available. A typical case may be a site agent working on a hospital with only limited experience of the large electrical, heating, or conditioning element of the work.

Although this adds to costs, it should be offset by any remedial work that may have had to be carried out on poor quality work resulting from lack of the correct supervision or inspection.

Communication

Drawings are often responsible for poor quality standards; they are often inaccurate, impractical and sometimes late, thus

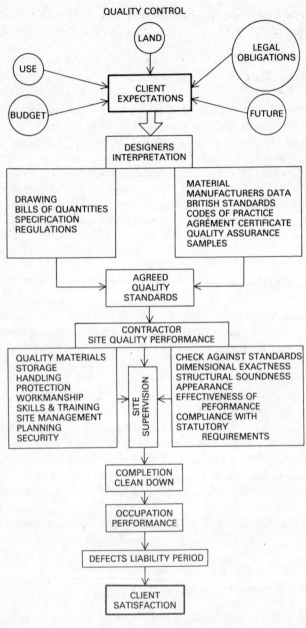

Fig. 44

creating not only the problem of progress and delayed work, but also costly time wasting. It also has a bad affect on motivation and therefore quality suffers.

Information must be co-ordinated, and a named person should be specified at the commencement of negotiations to be responsible. This is often the architect, which is advisable as it means that the contractor does not have to take the responsibility for design into his own hands.

Specifications

These play an important role in relating standards by written word, although they are costly to produce. Many sections are taken from old specifications and are irrelevant to current work. The National Building Specification (NBS) appears to be more realistic, from a recent survey of its use, achieving better results than the traditional form.

One of the problems with a specification, however, is the fact that the standards indicated still rely heavily on the interpretation of the reader; terms like 'to a high standard of workmanship' have little practical worth, and it is therefore important to lay down definite quantitive guidelines as far as possible, so that they can be checked, tested or measured with a degree of accuracy which is more positive than subjective.

Quality is obviously vital, but it must be related to progress and costs; there are many aspects that have to be clearly understood by all concerned in construction to ensure that required standards are achieved. *Training* plays an important part in assessing quality and a sound knowledge of technology and science play an integral part in the process.

15   Safety

Safety is the concern of all in the industry, from top management to the operative on the site. It also reflects itself at all levels when the cost of accidents are considered. It is generally overlooked that nationally over 20 million working days are lost through industrial accidents every year, which, in financial terms, is in the region of £100 million, and when this is compared with the average of 4 million working days lost each year through strikes, the significance of safety can easily be seen. The building industry safety record is not a good one, and an approximate breakdown of accidents is shown, Fig. 44. With the introduction of safety officers and far more concentrated education and training methods, improvements in this field are being made.

Cost of accidents

To the individual: some costs can easily be calculated when an operative has an accident, e.g. loss of wages, but far more items cannot be measured in cost, such as hardship brought about by reduced earnings, possible future earning capacity of the operative being permanently cut, and, of course, the human suffering and sorrows that no amount of compensation or benefits can erase.

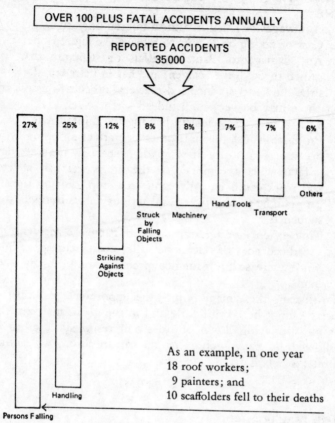

ACCIDENTS

OVER 100 PLUS FATAL ACCIDENTS ANNUALLY

REPORTED ACCIDENTS
35000

| 27% | 25% | 12% | 8% | 8% | 7% | 7% | 6% |

Persons Falling — Handling — Striking Against Objects — Struck by Falling Objects — Machinery — Hand Tools — Transport — Others

As an example, in one year
18 roof workers;
9 painters; and
10 scaffolders fell to their deaths

Fig. 45

Accident notification

The 'Reporting of Injuries, Diseases and Dangerous Occurrences Regulations 1985' (RIDDO) simplifies and extends the statutory requirements relating to fatalities, major injuries, lost-time accidents, dangerous occurences and industrial diseases. One of the new requirements is that all injuries at work which incapacitate for more than three days must be reported to the H & S Executive or the local Authority. This also applies to self-employed people and trainees. Failure to inform carries a maximum fine of £2000 (1987) or imprisonment.

To the employer: again some of the cost incurred can be calculated:

1 Any wages or benefits paid to injured operative.
2 Cost of taking on labour to replace the injured operative.
3 Any damage to plant, materials or structure that was caused through the accident and has to be repaired.

It is the hidden costs incurred that are difficult to assess and yet these may be very high indeed—items such as:

1 The working time lost in helping the injured man.
2 Production lost through other operatives stopping work because of curiosity or sympathy, especially if an accident is fatal, when the morale on the site is seriously affected and work can eventually come to a stop, causing the site programme to be affected and serious delays in operations sequence.
3 Time taken investigating and reporting the accident.
4 Overhead cost of telephone calls; cost incurred at head office increase in insurance premiums, and many such items.

Considering these many points, the importance of industrial safety cannot be taken lightly and to the individual firm this means the laying down of rules and responsibilities to be followed by each section of the organisation, an outline similar to details given.

Top management

Ensure that a clear safety policy is instigated, stating responsibility of all concerned, a system of adequate training, and all such items necessary to enable the satisfactory fulfilment of such a policy. (See Fig. 46.)

Site management: whose main duty must be to supervise and inspect all site operations ensuring that they can be performed in complete safety, taking as a guide the construction regulations, which will ensure that compliance with the law has been met. Compliance with regulations includes all necessary documentation.

Safety Policy

Under the terms of the Health and Safety at Work Act 1974 every employer has a duty to prepare a written statement of general policy with regards to health and safety of their employees while they are at work.

Most of the larger companies have designed a policy to suit their own concern, made up of sections, i.e. see Fig. 46.

---

1 GENERAL POLICY STATEMENT
. Stating the policy to safeguard the health, safety and welfare of all employees while at work together with, where practicable, people not in the firm's employment, so that they are not exposed to risk.

2 OPERATION OF THE POLICY
This section would indicate the responsibilities of all those belonging to the company from management down, and ensure that safety training and instructions are continually provided.

3 ORGANISATION AND ARRANGEMENTS
Specifies responsibility for all matters regarding health and safety, including provision for specific types of work, and the procedure for gathering and communicating new legislation, codes of practice, etc.

4 INFORMATION ARRANGEMENTS
This section is designed to monitor developments in safety, and indicate the type of information which should be collected, e.g. accident reports, developing training material, etc. together with inspection of places of work, plant transport and equipment, making sure at all times that Statutory Regulations are being observed.

Fig. 46

This is a very basic outline of a safety policy but it illustrates the wide area of activities covered in such a document.

Safety officer: now required by any firm employing more than 20 operatives. His is a specialist task to assist the site management in the execution of the firm's safety policy and to offer advice when necessary to ensure regulations covering safety are being maintained.

He is the co-ordinator between all levels on safety.

Generally also he is in charge of suitable safety training schemes.

Operatives: The operative's responsibility for safety is one basically of using common sense: for example, wearing safety helmets and other safety equipment; ensuring that the duties or operations that he performs will not cause injury to his fellow workers. Training plays an important role in this so that the operative can see the dangers likely to cause accidents.

It must always be remembered that every employee is under a statutory obligation to comply with all regulations and is liable to prosecution if he fails to do so.

Legislation

The present legislation is contained in the Health and Safety at Work etc. Act, 1974. The administration of the Act, although under the direct authority of the Secretary of State, has been built into the establishment of two bodies corporate, the Health and Safety Commission and the Health and Safety Executive.

The Health and Safety Commission consists of a chairman appointed by the Secretary of State and not less than six or more than nine other members, also appointed by the Secretary of State, although guide lines as to appointments are laid down in the Act.

The Health and Safety Executive consists of three persons one of whom is appointed by the Commission with the approval of the Secretary of State. This person becomes the director and after consultation and approval the Commission will also appoint the other two.

The Commission's role is with all relevant parts of the Act, including research, publication of results of research, besides putting forward proposals for the making of regulations. The Executive's duty is to perform the function as directed by the

Commission including the investigation or the making of special reports on any matter. It is the duty of the Executive to make adequate arrangements for the enforcement of the relevant statutory provisions, this they do by making it the duty of every local authority to make adequate arrangements for the enforcement within their area. Inspectors are appointed by the enforcing authority and in brief have the following powers:

1 Examine any site at any reasonable time.
2 Inspect any register, certificate, etc., which the Act and Regulations require to be kept and to take copies if necessary.
3 To seek out information from persons on site regarding information relevant to any examination or investigation.
4 If obstruction to the carrying out of his duties is anticipated the factory inspector may take with him a police officer.
5 If any substance may prove harmful to employees, samples may be taken.
6 May take measurements and photographs and make recording that he considers necessary for any examination or investigation.
7 Can require any person whom he has reasonable cause to be able to give any information relevant to any examination or investigation to answer questions as inspector thinks fit.

Two very important items an inspector has had added to his powers are:

1 Improvement Notices, which are served if in the inspector's opinion a person is
   a contravening one or more of the relevant statutory provisions, or
   b has contravened one or more of these provisions in circumstances that make it likely that the contravention will continue or be repeated.

   This notice informs the persons why the inspector considers provisions are not correct, making it necessary for the person to remedy the contravention.
2 Prohibition Notice is issued if in the opinion of the inspector the activities involved may risk serious personal injury. This takes effect if the inspector states that the risk of serious personal injury is imminent; in brief, work must stop. Persons being issued with notices may within a stated

period appeal to an industrial tribunal who have the power to cancel or affirm the notice or make modification as they think fit.

Existing Acts and Regulations are still applicable to building operations, although some sections of the Factories Act have been superseded by the Health and Safety at Work etc. Act. The main provisions for the building industry are to be found in:

Construction (Safety Health and Welfare) Regulations
        (General Provisions)
        (Health and Welfare)
        (Lifting Operations)
Lead Paint Regulations
Lead Paint Act (Section in the Factory Acts)
Offices, Shops and Railway Premises Act
Electricity Regulations
Protection to Eyes Regulations
Work in Compressed Air Regulations
Petroleum (Consolidation) Act
Petroleum Spirit (Motor Vehicles) Regulations
Explosives Act

To assist the factory inspector to carry out his duties, he will require various notifications from the builder, as follows:

1 Notice of building operations expected to last six weeks or more (Form 10).

2 Notice of Accident or Dangerous Occurrence (Form 43B). This requirement is for an accident resulting in death or absence from normal work for more than three days and other occurrences noted in instructions on the back of this form.

3 The Chief Inspectors of Explosives must be notified in the event of fire or explosion due to petroleum spirit or explosives (Back of Form 43B).

4 Notification required under the Lead Paint Act 1962 (now embodied in the Factory Acts) when persons are employed in painting buildings—no official form.

5 When certain diseases occur on site, notification to be made on Form 41 to factory inspector and appointed factory doctor.

6 Appointed factory doctor required to be notified within seven days of taking on young persons (i.e. under 18) into employment—no special form.

Prescribed notices

The notices quoted should be displayed either on the site or at the employer's yard, shop or office at which persons employed attend and in a place where they can be easily read.
1 Prescribed Abstract of the Factories Act (Form 3)
    On the notice should be shown the name and addresses of the following:
    a District and Superintendent Inspectors of Factories.
    b Appointed Factory Doctor.
    c Safety Supervisors.
2 Copies of the Construction Regulations.
3 The Woodworking Machinery Regulations (Form 988) if applicable.
4 The Electricity (Factories Act) Special Regulations, Form 954, if applicable.
5 The Lead Paint Regulations, Form 996.
    Displayed in workshop and paint store, and on any site where more than 12 persons are employed on painting operations.
6 Electric Shock Placard if applicable.

Prescribed registers

These are required to be available for inspection and use:
1 General Register for Building Operations (Form 36) which is kept in office or on site.
2 Prescribed Register for Reports Part 1 (Form 91 Part 1) which is kept on site and provides for reports on inspection of scaffolds and lifting appliances, excavations, tests for cranes and examination of passenger hoists.
3 Prescribed Register for Reports Part 2 (Form 91 Part 2), kept on site or in office, provides for reports on examination for lifting appliances, hoists, chains, ropes and lifting gear.
4 Prescribed Register of Persons employed in Painting Building, Form 92.
5 Certificate of tests and examinations of various lifting appliances must be kept in office or on site.

Causes of accidents

There are many causes and reasons why accidents happen and it is the duty of the site supervisory staff to know and to try to eliminate these to the best of their ability.

    a Weather is often a cause of accidents that would not otherwise happen. Rain, frost and snow produce dangerous walkways, slippery conditions, changes in ground conditions, often affecting excavations, hidden traps, or materials covered in snow. Wind can cause considerable damage, for it can easily blow over weak structures or loose sheet material, which can result in serious accidents.

Some other causes of accidents are as follows:

    b Poor communication in the giving of orders relating to safety measures.

    c Removing safety and protective guards.

    d Monotony in work—often forsaking safety precautions.

    e Trying to work with too much haste—often caused through badly worked out incentive scheme.

    f Defective plant and equipment.

    g Unsafe ladders.

    h Poor stacking of materials,

and many more reasons. All can be eliminated, and will be if everyone plays their part in accident prevention.

Insurance

Unfortunately accidents do occur and it is in the contractor's interest to take out insurance against claims due to accidents or death, not only for employees but also for Public Liability. The JCT Standard Form of Contract, if used, lays down in great detail necessary insurance cover. Briefly as regards to injury or deaths these are:

1   The contractor to indemnify the employer against proceedings in respect of personal injury or death of persons arising from carrying out works. If the employer or anyone who is responsible to the employer causes the accident by an act or neglect the employer is responsible.

2   The contractor must cover himself against injury or death

to persons outside his employment, i.e. third party insurance. This also covers sub-contractors.

3 Every employer must also take out employers' liability (compulsory insurance) under the Act of 1969. Copies of the certificate of insurance must be displayed prominently and suitably protected where they can be easily read by employees. For contractors this is usually at the employers' premises.

An inspector authorised by the Secretary of State can after reasonable notice ask to inspect current policies. Failure to comply with the Act may mean conviction and a heavy fine. The employer must insure for at least £2 million in respect of claims arising out of any one occurrence. In most cases this sum is available automatically with an 'approved policy' (one that is not subject to prohibited conditions). The policies must cover against liability for bodily injury or disease sustained by employees in course of employment.

Safety training

A most necessary attribute to any firm's or group's organisation is a suitable training scheme. This will most surely encourage all to regard safety in its proper light, which should reduce accidents and, in so doing, cut costs. Training is a continuous thing, starting at the early stages of a young operative's career, when it is important and easier to instil correct procedures and practices. Care must be taken, however, not to overlook progressive instruction as accidents often occur through 'old hands' getting careless—for the saying that 'Familiarity breeds contempt' holds very true and this is often the cause of accidents, especially in connection with machinery.

With most firms now having a Safety Officer, or even if only a share of a group Safety Officer, general training is more organised. Also with the development of safety training centres and various safety courses regularly being put on through the Construction Industry Training Board and others, knowledge on safety is being spread throughout the industry. The use of short lectures, films, handouts, posters and possible safety bonus schemes are all useful aids that are available to the Safety Officer. Whatever aid is used, it should be in conjunction with continuous education in safety and not just a 'flash in the pan' approach.

Lectures: should be of short duration, say one hour once a month. This will prevent operatives getting bored or loaded down with too much information at one time. It is best if a specific safety topic is tackled each time. This short lecture could be held in firm's time without much loss of progress.

Films or slides: could be shown in the lecture to create interest but whenever these are used, a brief summary or question and answer period should follow to ensure points shown are fully understood.

Handouts: can be obtained for putting in operatives' wage packets, and are useful to keep safety in his thoughts.

Safety bonus: a simple scheme can be designed so that a whole site can participate—for example: each man receives a pound standing bonus; if a few minor accidents occur, the operatives lose half—serious ones, the site loses complete bonus. This makes everyone aware of others' shortcomings, and an eye kept out for danger, as no one likes his pocket being hit.

Short courses: all personnel should be allowed to go on short courses as long as they are of a standard they can understand and will find helpful in their daily work. Safety is the concern of all, individuals, company and nationally; it is therefore important that all safety precautions should be adhered to and maintained.

Competent person: many items in respect to safety have to be inspected by a 'competent person'; this term in law has no meaning. It is therefore the employer's responsibility to decide who he considers competent. In the event of any legal proceedings, the employer may be called upon to satisfy a court as to the competency of the person, which may very well be related to the training he has received.

16 Work study

History

The development of 'work study' has been a constant process over the centuries, with man endeavouring to find and improve ways of doing his daily work. Many of these improvements did not cause great benefit on their own but have, with the passing of time, been collected and used by men with wide vision to produce results and labour conditions of a very high order.

Certain periods in time have also been responsible for creating, through the imagination of men, many advancements, for example the Industrial Revolution, when technical progress developed at unheard-of speed. Wars have also created situations which have produced ideas at a far quicker rate than would have been expected normally. It is therefore impossible to say who can claim to be the first man in this particular field, but generally, the honour is given to an American as the first to make a close study of the time for performing work; his name: Frederick Winslow Taylor (1856–1915). He first came to the fore as a lathe operator and gang boss at the Midvale Steel Works in Philadelphia, where he investigated problems of industrial organisation, one of which being the practice of 'soldiering', or taking things slowly so that rates for the job would be forced up. This resulted in the problem of doing a fair day's work for a fair day's pay, the problem being what constituted a fair day's work. Taylor tackled this and produced, after detailed analysis of jobs, 'standard times'. Unfortunately, the values used for the new system of payment were based on the performance of the best workers; this created discontent amongst the average and slower workers. Misunderstanding and controversy of his methods resulted for the rest of his life, but the work he did laid the foundation stone of work measurement.

Henry Lawrence Gantt (1861–1919) was a close colleague of Taylor, and developed a wages system that allowed even the slowest worker to benefit. Possibly Gantt's production programmes, for which he is remembered, were his biggest contribution. These are used in industry today and are known as Gantt Charts and show planning in graphical form in terms of times.

Frank (1869–1924) and Lillian Gilbreth have a close relationship with the building industry for he was a bricklayer in America in the mid-1880's. A study by Gilbreth into brick-laying techniques produced very progressive results. He studied and recorded the many ways that bricklayers had of laying bricks and reduced the movement involved, often from as many as 20 down to 5, with the result that these bricklayers, when re-trained, increased their output from an average 175 to 350 bricks per hour. Another of his forward-looking developments was a scaffold that could be continually adjusted so that the materials and wall being built were always at the most efficient height to reduce unnecessary bending and stretching. The methods of training and ideas he had were carried out when Gilbreth started his own building concern. Gilbreth's wife supplemented her husband's skill on technical aspects by relating the human needs of the operators to them as she was able to understand the problems of the worker. Mrs. Gilbreth was trained in psychology and sociology.

Many other famous people have, and are still, playing an important role in this development of the management 'tool' of combining industrial work with the human being, and the need will continue due to the competition to be found in our modern age.

The work-study engineer

The work-study engineer, when called in, often encounters a resentful atmosphere from management and worker alike. This feeling is generally caused by the lack of understanding of the study engineer's job. He is seen as someone out to 'check-up' on people—no one likes being watched and the thought of interference in our own little niche is very disconcerting. It is therefore the duty of management and supervisory staff at all levels to assist in any way possible, and by showing a helpful attitude, the men will follow their example. To do this, a supervisor must have an understanding of the operations and aims of work study; this is necessary to satisfy the workers as to the beneficial result of work study. What are these benefits?

To the organisation:
1   The reducing of cost resulting in bigger profits
2   More competitive tendering
3   Better control by management
4   Better use of resources
5   Collectively a better service to clients

To the supervisor:
1   More efficient planning
2   Better control
3   Fewer bonus target disagreements
4   Easier overall supervision
5   A more rewarding job

To the operative:
1   Better working conditions
2   Satisfactory bonus targets—steady high income
3   Less fatigue
4   Happier atmosphere and site relationship with foremen
5   More security—company successful

These will result through the aims of work study which basically are:
1   To increase productivity
2   To reduce costs
3   To increase profit margins
4   To make tasks fair and easier
5   To ensure security for operatives

6   To lay down standards and checks of estimates against actual
    costs
7   To lay down objective training
8   To reduce wastage
9   To cut out delays
10 To produce a more efficient organisation

Men should always be informed and kept informed during work study examination, and any orders or changes that have to be made must come from the foreman and not the work study man. It is also good practice, when installing a work study man on a site, to allocate him a suitable person, say a craft foreman or similar, to assist with technical queries.

Productivity

The ratio between output and input, or in other words the amount produced against the amount of the resources used in the course of production, is termed productivity. The resources in the construction industry are usually: land; materials; plant and equipment; manpower; or, as in most cases, a combination of all four. It generally falls to the management of an enterprise to try to achieve the greatest productivity possible, as only they have a true picture of all the resources available to them.

In order to achieve high productivity, management techniques have been evolved and developed. One of these 'tools' of management is known today as 'work study', a very broad term which embraces the whole sphere of investigations into production. Fig. 47 shows the basic aim of work study through the techniques of method study, and work measurement.

Most supervisors have, possibly without knowing it, practised work study in their everyday work, for any alteration that improves or increases efficiency is work study; this is generally a reflection of experience on the part of the supervisor knowing which is the best method. Work study can be used to do this in a scientific way.

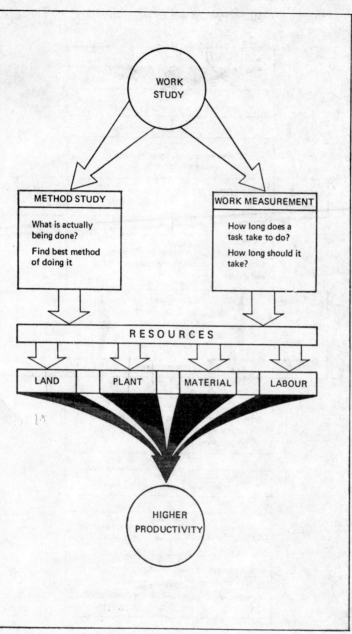

Fig. 47

OUTLINE PROCEDURE OF METHOD STUDY

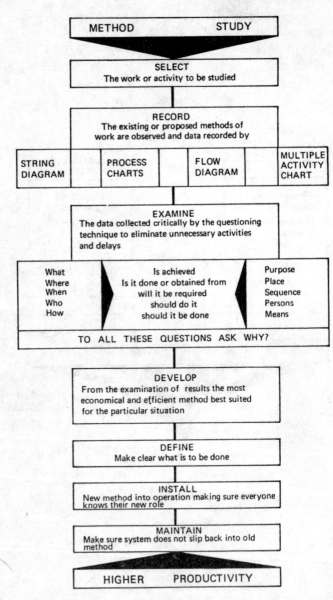

Fig. 48

Method study

This, in simple terms, is the study and recording of an existing or proposed way of doing work, and by careful and critical examination of the recordings produces an easier and more effective way of doing it. Fig. 48 shows the outline of the procedure with the various stages in order of study.

Techniques of method study

Scale models: These can be applied to various situations in the building industry; a typical application would be in the positioning of site huts on a scaled site plan; models or scaled cutouts can be used to find a satisfactory layout without a lot of pencilling and rubbing out every time an unsuitable position was chosen. Fig. 49.

String diagrams: Again using a scaled diagram placed on a suitable board, and with the aid of pins or tacks and a reel of cotton or similar, movements within a given area for a given length of time can be plotted and will show on completion, by unwinding the thread and measuring as per scale, the length of journey of plant, material or operative. It will also show crossing of paths and congestion bottlenecks. A number of paths can be plotted by using different coloured threads. Fig. 50.

Process chart: This form of recording information by the use of five easily understandable symbols can be seen in Fig. 51. The symbols represent the activities of a procedure or work method and show quite simply a visual step in a present or proposed method of work. Types of charts using these symbols are:

a Outline process charts
This chart is used, as the name suggests, to record a broad outline of a procedure, especially useful in showing the collection of materials or procedures to make up a whole

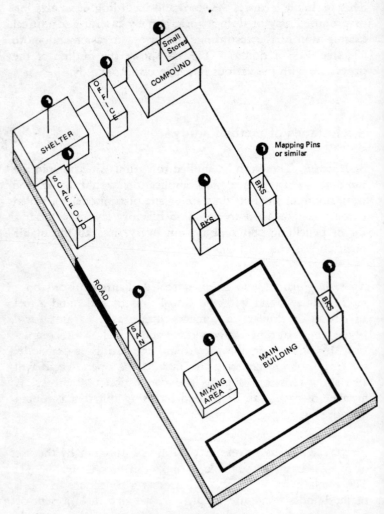

A SIMPLE MODEL APPLICATION – SITE LAYOUT

Fig. 49

STRING DIAGRAM

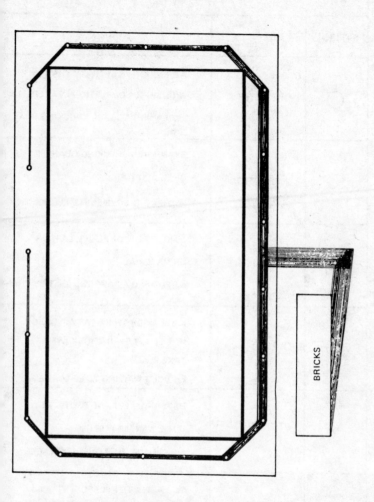

SHOWS PATH TAKEN BY LABOURER IN LOADING OUT
A SCAFFOLD WITH BRICKS. PINS ARE PLACED AT STACK
POINTS, THREAD IS TAKEN ROUND PIN EACH TIME IT IS PASSED

Fig. 50

RECORDING		
SYMBOL	ACTIVITY	MEANING
○	OPERATION	A PRODUCTIVE ACTIVITY THAT WILL BRING COMPLETION NEARER e.g. LAYING A BRICK, KNOCKING IN A NAIL
□	INSPECTION	EXAMINING—CHECKING QUANTITY OR QUALITY e.g. MEASURING OPENING; PLUMBING UP
▷	TRANSPORT	A MOVEMENT OF PLANT, LABOUR OR MATERIAL e.g. PUSHING A BARROW; HOIST BY CRANE
D	DELAY	A DELAY OR TEMPORARY STORAGE WHEN NEXT OPERATION CANNOT TAKE PLACE e.g. HOIST WAITING TO BE LOADED
▽	STORAGE	MATERIAL KEPT AND PROTECTED UNTIL WANTED FOR USE e.g. CEMENT IN SILO; BRICKSTACKS
◙	COMBINED ACTIVITY	TWO ACTIVITIES PERFORMED AT THE SAME TIME OR BY SAME OPERATOR e.g. LOADING AGGREGATE INTO WEIGH BATCHER

Fig. 54

unit as in Fig. 52; only operations and inspections are shown.

b Flow process charts

These are used to develop and expand the problems with the aid of the other symbols—transport, delays, storage— a typical example is shown in Fig. 53. The chart can be made to record the movements of men or material.

c Flow diagrams

Drawn on a scaled plan and used as a supplement of the flow process chart, show the path of material or operatives by the use of a line joining the process symbols together. A simple example can be seen in Fig. 54 and shows how it could supplement Fig. 53.

Multiple activity chart: This is a very efficient way of recording information where it is necessary to relate one subject or activity with others; it has proved to be a very useful aid to work study in the construction industry due to the many operations that are performed by teams of men in gangs. The chart itself is produced in bar form against a time scale and generally the use of an ordinary wristwatch is sufficient to record times. An example is shown in Fig. 55. This shows a very simple application, by increasing labour force by one. 20% extra concrete is produced with a possible saving due to transport being used more efficiently and placers being used to full capacity. This would have to be assessed from further study, as against the cost of an extra labourer.

Activity sampling: The technique of activity sampling allows a statistical method of study to be used on a site by an observer with little or no continuous close contact with the operatives or plant being studied. Briefly what occurs is that the observer 'snaps' the object of the study at a frequency that has been established before commencement of study to ensure correct degree of accuracy is achieved. These 'snaps' should be made as randomly as possible over a stated time scale which may be a day, or more practically, several days. It is reasonable to suppose that a greater number of snaps over a longer period of time will make the sample much more representative.

A practical example would be recording the times a piece of

173

OUTLINE PROCESS CHART

STUDY No 17   CHARTED BY A.N.Other   DATE 06th Mar 79
SHEET No 2
OPERATION DESCRIPTION *Batching and Mixing Concrete*
CHART BEGINS *Raw Materials*
CHART ENDS *Mixed Materials (Concrete) Discharge*
~~MAN~~/MATERIAL/~~EQUIPMENT~~

WATER IN TANK	CEMENT IN SILO	FINE AGGREGATE STOCK PILE	COURSE AGGREGATE STOCK PILE

5-4 Water to Mixer. Check Gauge

3-5 Fill & Weigh

2 Load Hopper

1 Load Hopper

2 Check Weight

1 Check Weight

4 Charge Drum

6 Mixing Time

5 Check Mix

7 Discharge

SUMMARY.   OPERATIONS 7
           INSPECTIONS 5

Fig. 52

FLOW PROCESS CHART

STUDY No. **23**   CHARTED BY **A.N. Other**   DATE **06th Mar 79**

SHEET No. **3**

OPERATION DESCRIPTION **Off loading of sinks into storage compound**

MAN/~~MATERIALS~~/~~EQUIPMENT~~

~~RESENT~~/~~PROPOSED METHOD~~

DESCRIPTION	Quantity	Distance	Time	Operation	Inspection	Transport	Storage	Delay	REMARKS
1 labourer walks from building	36	60m	1m	○	□	▷	▽	D	1 Labourer
Collects barrow		10m	0·10	○	□	▷	▽	D	
Unload from lorry	3		0·75	○	□	▷	▽	D	By hand
Takes to stores	3	50m	1·30	○	□	▷	▽	D	In barrow
Checked	3		0·30	○	□	▷	▽	D	By storeman
Unload & stack in stores	3		0·60	○	□	▷	▽	D	
Return to lorry		20m	0·20	○	□	▷	▽	D	
Repeat 3/9 11 times	33	770	34·65	○	□	▷	▽	D	
Return barrow		10m	0·10	○	□	▷	▽	D	
Walk back to building		60m	1m	○	□	▷	▽	D	
				○	□	▷	▽	D	
				○	□	▷	▽	D	
				○	□	▷	▽	D	
				○	□	▷	▽	D	
				○	□	▷	▽	D	
				○	□	▷	▽	D	
				○	□	▷	▽	D	
				○	□	▷	▽	D	
SUMMARY: PRESENT	36	980 m	40 m	24	12	24	–	4	
PROPOSED									
SAVING									

Fig. 53

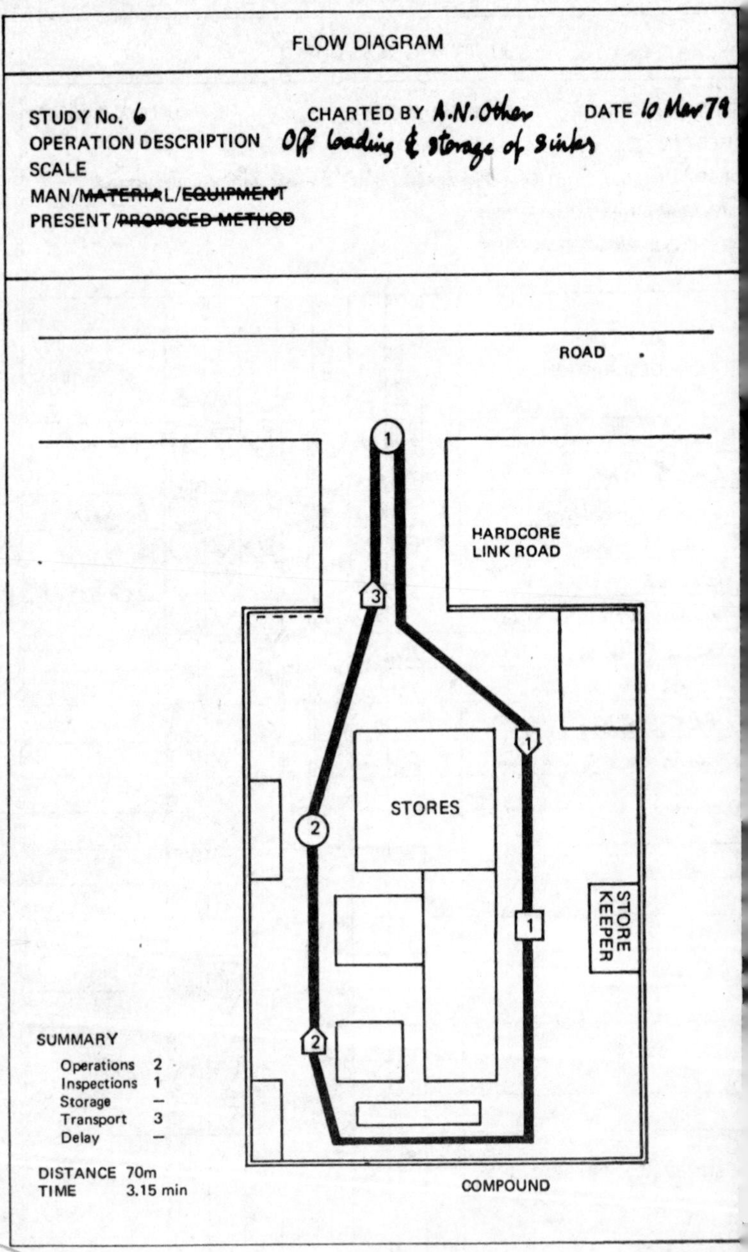

FLOW DIAGRAM

STUDY No. 6   CHARTED BY A.N.Other   DATE 10 Mar 79
OPERATION DESCRIPTION Off loading & storage of sinks
SCALE
MAN/~~MATERIAL/EQUIPMENT~~
PRESENT/~~PROPOSED METHOD~~

ROAD

HARDCORE
LINK ROAD

STORES

STORE
KEEPER

SUMMARY

Operations	2
Inspections	1
Storage	—
Transport	3
Delay	—

DISTANCE 70m
TIME   3.15 min

COMPOUND

Fig. 54

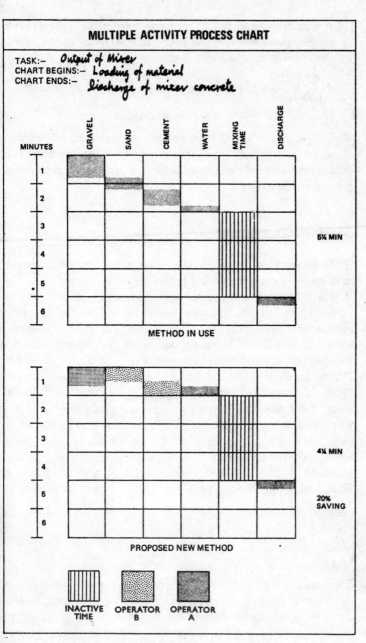

Fig. 55

plant was operating. If 120 readings were made over a period of say five days, and if it was noted that 40 times the plant was idle, it is easily worked out that

$$\frac{40}{120} \times \frac{100}{1} = 33\frac{1}{3}\% \text{ of the time the plant is idle, making}$$

$$66\frac{2}{3}\% \text{ it was in operation}$$

This type of work measurement has several uses, but as with all statistical methods great care should be taken to ensure that data is read correctly and that the degree of accuracy is adequate for the study in hand.

Work measurement

This is used to establish the time taken to do a specific task and is a complementary technique of Method Study. The aim is to cut ineffective time, with the worker's effort, production and conditions of work being related to the necessary relaxations allowances to ensure that undue fatigue does not occur. It is believed by many that work measurement cannot be operated on the building site because of the many variables that occur such as the changes in climatic conditions, different types of materials, varying conditions of work and so on. This is not the case, for there are many occasions when, by work measuring, operatives' ineffective time can be shown to exist and a new system developed; also the time studied operations provide an excellent base for competitive estimating, planning, costing and the forming of sound incentive schemes. The outline procedure of work measurement is shown in Fig. 56.

Time study

This can be carried out quite often by the use of an ordinary wristwatch or stop watch and is possibly the most useful form of work measurement used in the building industry; it can be used on both single operatives or gangs, as required. Stages in time study are:

178

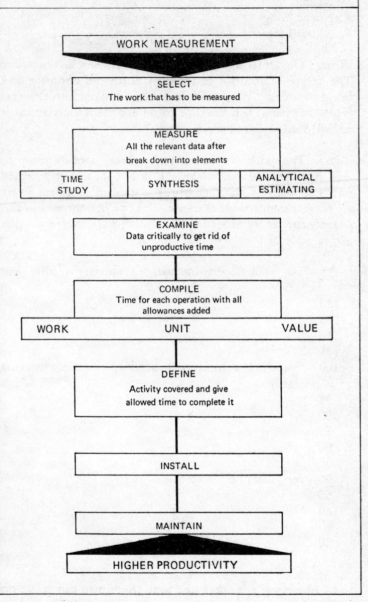

Fig. 56

1   Timing
2   Rating
3   Normalising
4   Allowances.

Timing: The actual taking of the time to complete an operation. The degree of accuracy in timing will depend upon the task being recorded. The operations are first broken into elements (operating parts) and the time to do each separate element is recorded on prepared sheets, Fig. 57.

Rating: This is the allocating to a worker a relationship to 'Mr. Average' and as this 'average' does not exist, an accepted rate is applied; or, in other words, a standard is given. There are several standard scales in use. The one mainly used in this country and recommended by the British Standards Institution uses a scale of 0 to 100.

125—fast, skilled performance of necessary quality and accuracy
120
115
110
105
100—represents a motivated qualified worker, working normally
95
90
85
80
75
70—very slow, no interest
65
60
55
50

By observing a worker, the time study man can relate his effort to the scale, so that slower workers will be given a

TIME STUDY FORM

CONTRACT	STUDY NO	DATE
OPERATION DESCRIPTION	SHEET NO	

OPERATIVES PLANT/EQUIPMENT CONDITIONS	TIME START TIME FINISH LAPSED TIME OBSERVED TIME DIFFERENCE
	CHECKED

Element Description	R.	WR	O.T.	B.T.	Element Description	R	WR	O.T.	B.T

NB. R=RATING WR=WATCH READING OT=OBSERVED TIME BT=BASIC TIME

Fig. 57

figure below 100 and the skilled efficient man can achieve over 100. Considerable skill is required to give a true rating: for example, a bricklayer knocking down a brick may look and sound very effective, whereas another bricklayer may calculate the correct thickness of mortar bed so that the bricks only require placing. This is the effectiveness of a worker and will relate to his skill in doing the work.

Normalising: This is the time that in the judgement of the work-study man an element should be performed. To obtain this normalised time, a simple formula is used:

$$\frac{\text{observed time} \times \text{observed rating}}{\text{standard rating}}$$

This means that every element is converted by multiplying the appropriate rating to a normal time.
Example: if an element took 1·4 min and was given a rating of 90, the basic time would be:

$$\frac{1·4 \times 90}{100} \text{ (scale 0/100)}$$
$$\text{Basic time} = 1·26$$

This time of 1·26 min represents the time it would take if the operator was working at a normal rate, 'Mr. Average'.

Allowances: These allowances can be divided into two groups:
1   Process allowances
2   Rest allowances
A process or unavoidable delay allowance is given so that the worker will not lose earnings due to an enforced delay over which he has no control, e.g. the cleaning of tools or plant.
Rest allowances vary with every performance and condition of work, and are added so that a worker can keep physically and psychologically fit to perform an operation for an allotted time. The allowance is in fact sometimes termed 'fatigue allowance'.

Items taken into account are as follows:
Basic 9% minimum.
This is the minimum requirement for the personal needs of every person regardless of work.

Variable rest allowance
Position of body, 0–7%.
This will vary from the sitting in a comfortable chair to working in a cramped position understairs or similar.

Conditions, up to 15%.
Include such items as: noise, bad light, dust, heat, cold and water. The difference in conditions are numerous and can range from working in a hot boiler house to excavating in a deep muddy wet trench.

Mental Strain, up to 8%.
This relates to the concentration that needs to be applied to a piece of work or a task. For example, a tower crane driver positioning heavy loads, or a painter doing detail design.

Manual effort, up to 20%.
A condition very apt in the building industry where a great amount of muscular force is used from demolition to manual unloading and carrying materials. The allowance is calculated by the work study man in the weight that is lifted, pulled or pushed.

Monotony, up to 4%.
Generally not very serious in the building industry as most jobs last for a comparatively short period; there are occasions, however, such as hammer and chisel work in chasing for services.

By adding up the related percentage given to a task, a total rest allowance is obtained; if this figure is high, consideration should be given to reducing it as it would appear that the task to be carried out by the worker is too arduous; about 40% is an approximate maximum generally accepted.

Contingency allowance

This is a special allowance applied to various tasks that have to be carried out by the worker who will not be doing a productive job: as an example, a joiner sharpening saw or chisels. These allowances should not exceed 5% and should only be applied to justifiable cases.

Allowed Time
A final time allowance for a task can now be ascertained by the simple calculation shown.

Example: Basic Time: 1·26 min
            Rest allowance: 30%
            Contingency allowance: 2%

Standard time = Basic time $\times \dfrac{100 + \text{Total \% allowance}}{100}$

$$= 1·26 \times \frac{132}{100} = 1·66 \text{ (say } 1·7\text{) standard minutes}$$

Make up of Allowed Time:

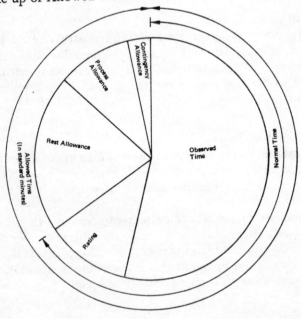

Use of work measurement

The resulting times obtained are the times that should produce the highest productivity for operatives. It is therefore of advantageous use in the preparing of estimates for tendering purposes as the times and prices are now carefully calculated and not guessed. Also with studied times the application to planning and all forms of control is invaluable to set standards.

Synthesis

This is the use of previously obtained element times that may be recorded in some form of library. When time for an operation is required, the task is broken down into elements and the times for these elements are collected from the library.

Analytical estimating

This is similar to synthesis, but where element times are not in the library they are estimated; this will mean that any small items of a task will have to be assessed and not the whole. Often used in repair and maintenance, it requires considerable skill to do satisfactorily.

17 Incentive scheme

Introduction to Work Study by the International Labour Office states that an incentive scheme is:

Any system of remuneration in which the amount earned is dependent on the results obtained, thereby offering the employee an incentive to achieve better results.

The National Working Rules State:
Productivity—Incentive Schemes and/or Productivity Agreements.
Incentive schemes and/or productivity agreements should, wherever practicable, be operated by employers on jobs and in shops and factories in order:

a to increase productivity and reduce costs

b to enable operatives to increase their earnings by increased effort

Incentive schemes shall be drawn up in accordance with the Principles agreed and published by the Council.

In the same statement of General Principles there are set out provisions governing the making between employers and operatives of productivity agreements which take into account technological change and have as their objectives (a) the achievement of a higher rate of productivity through the more effective use of labour, and (b) the provision of opportunity for higher earnings.

There are two types of incentive: financial and non-financial, which can be met in all walks of life, from receiving interest from a bank or a building society (incentive to save), to receiving a free gift with a purchase (incentive to buy). To the worker, the main incentive is generally one of an increase in his earnings and is the firm's main method of motivating the operatives, or in other words, creating a situation that encourages a single person or a group of persons to work harder and produce more; in so doing, it should also be possible to reduce the cost of the unit being made, basically because more units are being produced with the same cost of overheads as before increase in production. An outline of the basic organisation required for the operation of a bonus scheme is shown, Fig. 58.

The W.R.A. states the objects of incentives in the building industry:

1   To increase efficiency by reducing cost of building
2   to increase individual and collective production
3   to provide opportunity for increasing earnings

If these objects are achieved, it follows that in any proper incentive scheme, payments should be strictly related to production.

Financial incentives

Profit sharing: Often used in smaller concerns and as a method of holding labour for a given period, say 6 to 12 months, with the incentive being a reasonable cash settlement at the end of this period, related to the success and profits of the company. A big disadvantage in this system is that 'old soldier' types will benefit from the efforts of the hard worker and a hard-working productive supervisor will have to carry supervisors on other sites who are not so industrious.

Hourly plus rate: Used by many companies to attract labour, especially if in short supply. It may also be used when the quality of the work in hand is more important than the quantity produced. The main disadvantage with this scheme is that the employer has no guarantee that the extra rates are producing the required extra increase in production.

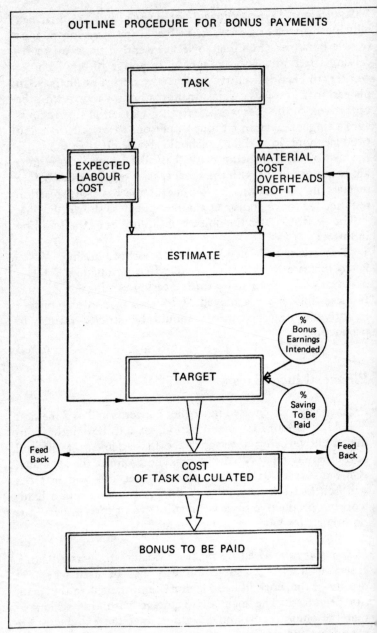

Fig. 58

Bonus schemes: The most common type of financial incentive used in the industry, for it does relate earnings to effort if correctly applied.

There are various schemes used to obtain the required results and these should be carefully studied in relation to the type of project on hand. Most of the schemes work on a target basis expressed either in terms of time or value. Basic principles should be applied on all these bonus schemes to ensure that it runs successfully:

a Scheme should be simple to understand so that operatives can assess increased earning, remembering to keep tasks small.

b Output and quality of work based on the average operative working under average conditions—work measurement can be applied here.

c The percentage saving achieved on a target to be paid out to the operatives must be agreed before commencement of scheme. This can rate from an operative receiving 100% of saving to only 50%.

d Targets should be set down in writing.

e Targets should not be altered unless upwards; it may possibly be better, therefore, to start low and increase bonus earnings if necessary.

f Payments should be made at regular intervals, e.g. once a week.

g Losses of one week should not be deducted by gains in other weeks. Once a scheme is in operation, great care must be exercised to ensure that the incentive scheme does not produce any running down of standards in the following:

a Safety of the operatives as laid down in the Construction Regulations.

b The material waste of normal working is not increased.

c The required standards of workmanship are still operational—payments should not be made for poor work.

d Plant used in process is used efficiently and for the purpose it was intended.

e The training of apprentices should still continue to a satisfactory standard.

To help control standards, it is generally agreed that site

supervisory staff whose duties are purely supervisory should not be included in schemes.

A simple method of recording and calculating a bonus scheme suitable for house development is shown as follows.

SCHEDULE OF BONUS RECORDINGS						
Labour Gang A and B					Week ending 21 March 1979	
Item	No. of units completed	Target hours each	Total target hours	Man hours worked	Hours saved	Bonus hours 75% saving
First floor joist	12	5	60	44	16	12
Boarding Ground Floor	8	6	48	36	12	9
Roof Carcass	4	30	120	100	20	15
Total	—	—	228	180	48	36

N.B. It is recommended that the target should be such that the average worker should be able to earn 20% bonus under normal conditions.

Non-financial incentives

It is not always necessary to give extra payments to motivate people. This can be seen in all industries where in certain areas the wages are not as high as local competitors, but labour relations and working conditions are good. These conditions are many but fall under the general heading of non-financial incentives. Some of these are as follows:

Working Conditions: the general conditions in which a worker has to perform his allotted tasks: On the building site it can generally be seen by the standard of the site layout and offices, materials and the general appearance of the job. Winter building is always a problem in the industry, when men leave cold wet sites for the warmth of factories. This is now being overcome by projects being totally covered.

Promotion: always an incentive to the ambitious man is a

chance of getting to the top, and to achieve this promotion a man will put extra thought and effort into his daily work.

Security: possibly one of the best known and sought-after non-financial incentives, especially when operatives have family commitments, also when unemployment is prevalent. The firm that offers continuity of work will always attract a steady labour force.

Safety: unfortunately this is not one of the industry's best selling points and many workers will work more confidently and willingly if they feel safe getting to and from their place of work and whilst they are carrying out their productive operations.

High class work: will often act as an incentive to men for the honour of working on a building of, for example, national importance.

Other non-financial incentives are:
a Social activities
b Pension schemes
c Extra holidays
d Provision of meals
e Training
    and many more.

Guaranteed Minimum Bonus Payment and Supplement

The WRA states: As from the commencement of employment an operative shall be entitled to receive a guaranteed minimum bonus payment subject to conditions laid down in the WRA with the sum payable set off against all other bonus payments or extra payments other than those prescribed in the agreement, in no case additional to them.

Also under the 1975 Terms of Settlement all operatives were entitled to receive a Joint Board Supplement subject also to the conditions set out in the NWRA.

18 Communications

Communication is as old as time itself; man has always wished to talk and join with others. As the world has developed and in fact been made 'smaller' through the means of communication, so the systems used have become more sophisticated and complex. We have only to consider today the fantastic step forward made possible by satellite, television and such. Communication is of vital importance in the controlling, co-ordinating and motivating of people, and all through history it has played its part; for example: Nelson's immortal message at Trafalgar, a message that stirred many to great efforts; Churchill's memorable speeches during the second great war, when millions received hope and encouragement through the medium of radio.

Just as important as communication is that the person or people receiving it should be able to understand, for errors can be costly, as we can again see from history when we consider the Charge of the Light Brigade, a famous battle but a disaster, due to a misunderstanding in communications. The importance of good communication can readily be seen and is just as important for the survival of a company as it is nationally.

Communication in the building industry generally takes one of three forms:

1 Oral instructions

2 Written instructions
3 Drawn details

Whichever method is adopted, certain considerations relate to all. They should be:

Precise: the communication should be clear, straightforward, as simple as possible, accurate in every detail, well thought out.

Not too long: so that parts may be forgotten, misunderstood or misinterpreted.

Definite: so that no doubt is left as to what the message means; should not be changed once it has been given.

Situation: communications should suit both the situation and the person receiving it.

Oral instructions—orders

These should be given in a manner that reflects efficiency and enthusiasm. The posture should be upright but relaxed; the speech clear, calm, yet commanding. The verbal instruction should be given directly to the person concerned, otherwise the message could become distorted with passing on. The 'face to face' talk will generally have far more success than the written word; it will also enable questions to be asked and queries raised without delay. It should always be remembered that most people prefer to be asked to do something rather than be told or ordered. The person being ordered to carry out the task may have no option but to do it, yet he will more willingly and possibly more efficiently do it if asked to so do in a courteous and friendly tone.

The telephone

Oral messages or instructions are often given and received over the telephone. The correct use of this instrument is important as it often reflects an image to the person receiving the call, who may never have met the caller. The person using the 'phone should speak clearly, without shouting, speaking in a normal conversational tone, giving the listener time to reply to any matters arising.

Before dialling: it is always a good idea, before using the 'phone, to think about the information required. A few notes can be made of relevant details, as this prevents points being missed and saves the necessity of phoning again.

When receiving: a courteous greeting, e.g. Good morning, or Good afternoon, should precede the name of the company or number, necessary so that the person calling knows he has the correct line, then one's own name or department should be added so that it is known to whom the caller is speaking. A pencil and pad should be made handy for noting down any messages. These messages should always be read back to the speaker to ensure correctness. If a promise to ring back or pass on a message is given, then this should be done without fail.

It is good practice to keep a list of names, telephone numbers of firms, and individuals associated with the organisation, for easy reference. A telephone register for the recording of incoming and outgoings calls, showing clearly the date, firm's name, person's name and brief details of calls is an efficient means of keeping record of calls and may help in any disputes arising, e.g. order given over the 'phone.

When receiving messages on behalf of someone else they should be written on a memo pad or similar and the person concerned contacted as soon as possible; when the message has been passed on, the signature of the recipient should be obtained in the register.

Written instructions

The written word comes into many forms of communication found in the building industry—letters, reports, minutes of meetings, general documentation.

Paperwork generally has increased a great deal in modern industry, mainly due to the much more careful control required to run a successful and efficient business, for documentation provides the necessary links in the chain of administration. All members of an organisation, large or small, should be made fully aware of the importance of correct paperwork in all its aspects, from letter writing, that projects a firm's image, to site documents that can be used for future reference and control.

Memoranda (memo)

Printed forms used to convey written or typed messages within the organisation's own structure, e.g. planning department to estimator. A typical memo heading would be as shown:

```
+------------------------------------------------------------+
|                        A.N. OTHER LTD.                     |
|                            MEMO                            |
|                                                            |
|  From.............................      To...............................|
|                                                            |
|                                         Date:              |
|                                                            |
|  Subject                                                   |
|                                                            |
|                                    (Space for message)     |
|                                                            |
|                                    Signature of sender     |
+------------------------------------------------------------+
```

Letters

There are many occasions and circumstances which require the writing of letters to convey messages to people outside the structure of the company.

General considerations

1 Use organisation's official notepaper
2 Take care over English, punctuation and spelling
3 Ensure any references are quoted
4 All letters should be dated and signed
A typical letter is shown, to illustrate an example of layout.

A.N. OTHER LTD.
BUILDING CONTRACTORS
1762 BLOCK ROAD
REDHILL,
WARWICKSHIRE

Tel. 041 6996 341

Our Ref. CB/JS

Your Ref. 146/7/MJ

26th September, 1979.

J.K. Black Ltd.,
Builders Merchants,
Stanford Green,
NORTHINGTON.

Dear Sir,

Order 1672/69

With reference to the above order, we have to advise you that four of the plastic light fittings were found to be broken upon receipt of the order on the 25th September, 1979.

We shall be pleased if you will replace these as soon as possible.

Yours faithfully,

J. King

J. KING,
Manager.

Example 1

1 Letter on firm's headed paper, giving full name, address and telephone number.
2 Date: should be typed in full, just below the letter heading.
3 Reference: generally on the left-hand side, opposite date.
4 Addressee's name and address: typed just below the reference.
5 Subject: heading, in centre of sheet and usually underlined.
6 Salutation: always on left-hand side of letter and generally either:
   a Dear Sir(s) or Madam—(strictly business)
   b Dear Mr. Black—(less formal)
   Dear Mrs. Smith
   Dear Ms. Evans
   Dear Miss Jones
   c Dear Mr. Black
   Dear Black—(more personal)
   Dear John
7 Body of letter: if of a business nature, should be short and to the point, care being taken not to omit relevant information.
8 Complimentary close: toward the lower right-hand side of sheet, should be complementary to salutation:
   a Yours faithfully—(order as for salutation)
   b Yours truly,
   c Yours sincerely.
   Only the first word is capitalised in closing.
9 Signature and designation: Person sending letter signs—name typed, with position or rank in firm under.

Reports

A report is used to relate information back to someone, such as investigation, something witnessed, or suggested ideas. The main object of a report should be to disclose facts for recommendation without the 'padding' as used by a fiction writer. Superfluous wording should be avoided, but care must be taken to ensure readers fully understand what the report is meant to convey. A report can be divided basically into five sections:
1 Headings

2 Introduction to report, its purpose
3 The body of the report—use of sub-headings and dividing information into sections should be the aim (useful for quick reference)
4 A conclusion, or summary, with any possible recommendations
5 Any data, drawings, tables, etc.

Example 2:

Contract: School   Contract No. 196/69
         Junction Road,
         Blackwater.   Date: 16th March, 1979.
To: J. A. Black, Safety Officer.
From: P. P. White, Site Agent.

Unsatisfactory Safety Record

There appears to be little thought given to safety training on site, with the result that accident figures are high.

Action already taken:
    a A safety committee comprising of three representatives from the men and myself has been set up to try to resolve the problems existing on site.
    Points raised:
    (i)   Not sufficient trained scaffolders
    (ii)   No site training for operatives
    (iii) Possibility of safety bonus
    b Safety Posters have been displayed around site.
    c Accident indicator board has been made to show types and numbers of accidents that take place on site.

Recommendations:
I suggest the following action be taken:
    a Careful consideration be given to the introduction of a safety bonus.
    b Arrangements made for operatives to attend short lectures on safety in the firm's time—one hour's duration would be sufficient once a month on a topic.
    c A safety handbook to be prepared by firm, each operative receiving a copy. Highlighting accident black spots and possible dangers.

It is appreciated that the recommendations above will cost money, but when the cost of the accidents is taken into account and the time lost, it is clear that this would pay for itself in a very short time.

Signed:
P. P. WHITE

Note: information regarding the cost of accidents and the time lost on site could be handed in with report in the form of graph or statistics.

Example 2 shows a typical example of a detailed report concerning a specified item whilst Example 3 shows a standard form of report which would be used at regular intervals to provide basic details of progress; other types of printed report can also be used to cover such items as accident or similar.

Notice boards

A useful means of conveying information, if placed in a prominent position where everyone can see it, e.g. by the time-clock or in the canteen—not under the coat pegs.

Handbooks and regulations

Copies of such items as the Construction Regulations, Building Regulations and the Working Rule Agreement should be kept on site in a position where they can be reached easily for points of reference, not locked away in a drawer or buried under drawings.

Site records

Many site records have to be made out each week regarding labour, plant and materials. It is important to remember that these documents have a specific use and must, therefore,

A.N. Other Ltd.,
1762, Block Rd.,
Redhill, Warwicks

A.1691

WEEKLY REPORT

Contract ....

Week ending..

Sub-contractors on Site .... Trade .... Labour Force

....

....

....

....

....

Number of men on site in our employ:

Bricklayer .... Bar bender ....

Bricklayer Labourer........................... Machine driver ....

Labourer ....

Carpenter ....

Painter ....

shall require the following:

Materials as requisition Note No .... by ....

.... by ....

.... by ....

Information from .... by ....

Repairs to plant as requested .... by ....

Plant .... by ....

Drawings .... by ....

I shall be ready for the following sub-contractors ....

.... on ....

.... on ....

.... on ....

.... on ....

.... on ....

The job has been visited during the week by ....

....

Have all accidents been reported? ....

Have all instructions been recorded in the instruction book? ....

....

Man hours lost during week:

Craftsmen .... Labourers .... Reason ....

The job is ahead of/behind schedule by....... weeks for the following reason

....

Other information considered to be of importance ....

....

....

Foreman's Signature ....

NOTE:— This weekly report should be sent in to the Office with the wages sheets each week, in the special envelope provided to be received not later than the Tuesday morning following, and it has been instituted with a view to assisting the Foreman in the progress of the work.

Example 3

always be filled in accurately and when required, whether daily, weekly or monthly. These documents are dealt with separately in various chapters.

The building site daily diary

This has often proved to be a most useful document and if well kept, with the right type of information recorded in it, its value in matters of dispute later in the project cannot be over-emphasised. Even long after project completion, it has often been of utmost importance in cases that have gone to court. The main objective of the site diary should be to record events and information that do not warrant special records being kept. Different supervisors will use the diary in different ways, placing more importance on certain items than others. As a general rule, items which might well be recorded could be:

1 Telephone promises from sub-contractors, suppliers, etc.
2 Verbal instructions from architect
3 Visits to site by building owner, architect, quantity surveyor, factory inspector, etc.
4 Details of weather conditions, especially in winter months
5 Delays in programme due to late delivery of materials, late start of sub-contractors, etc.
6 Verbal instructions from head office or contracts manager
7 Any matters of unusual occurrence, decisions or actions it is felt should be recorded.

Drawing

An important feature here is that of the architect's drawing and details to enable the project to be built. It will be realised therefore that in this form of communication great care is required to be practised by the architect to present the drawing in such a fashion that it cannot be misread or mistakes made, whilst on the other hand the site staff have to have knowledge to understand drawings in order to interpret them correctly.

A proper plan chest should be used for storing these plans and details, to ensure that they do not become lost and badly damaged. Any corrections should be clearly marked on drawings and not carried round in the foreman's head.

Another very useful form of communication is a sketch, and it will often be found to be far more advantageous than long written documents. The art of sketching is, therefore, one which the supervisor should practise, not to become an accomplished artist but to be able to show in a three-dimensional picture what is required in a simple fashion.

The master plan

This is found in the site office and is also a means of communication, for, as can be seen from the ones in the chapter on planning, much information can be obtained from it, besides what is added to keep it up to date and useful.

Conclusion

There are many and varied ways of communication but all must rely on the same basic principle: can the receiver understand the sender? This is not quite so simple as it may first appear for certain barriers exist that can cause the breakdown of this understanding.

1 The physical distance between sender and receiver
2 The method of communication that may suit one person may not have any effect on another
3 Superiors who simply dictate orders and are not prepared to listen
4 Fear of losing a job if disagreement over communication
5 Hurried instructions or details given without full thought or understanding as to the finished result
6 The way the spoken word is given: the speaker may understand what he is saying, but can the receiver?

All communications must be two-way.

Meetings

The smooth and efficient running of any organisation relies upon information being exchanged at all levels. Meetings are arranged to ensure that parties concerned or interested in events can meet and discuss points.

To be effective, meetings must be run correctly, in an orderly yet friendly manner, without any unnecessary time-wasting.

Chairman: to produce results, meetings should be run on organised lines. This means first electing a chairman, whose role is to control the meeting, generally summarising so that points raised during discussion are quite clear to everyone, to see that the agenda is adhered to, and open and close the meeting. At site meetings, the chairman is generally either the architect or a senior site supervisor. For example, at an architect's site meeting, the architect would take the chair; at a contractors' site meeting, the agent or similar would take control.

Secretary: The secretary's duties are primarily to take down items raised at the meeting, generally preparing all the arrangements including the agenda, and getting out and circulating the minutes. The secretary is usually a member of the architect's clerical staff, the site clerk, or similar, depending upon who calls the meeting.

Persons present: This will of course vary with the reason for holding the meeting, for only people who are interested or can play an active part in the proceedings should be called.

The Agenda: It is vital that meetings are carried out in a logical sequence so that nothing is left out. To achieve this, an agenda is sent to all members required to be present. This ensures that the meeting will be carried out in sequence and will also enable people to collect data or check on items stated on the agenda before meetings take place. This may result in the saving of considerable time and trouble. A typical agenda is illustrated:

AGENDA of meeting of Safety Committee, to be held at . . . .
(Place) on . . . . (date) at . . . . (time).
1 Apologies for absence
2 Minutes of the last meeting
3 Matters arising out of these minutes
4 Any items carried over from last meeting
5 Main business of present meeting (these would be listed)
6 Any other business
7 Possible date and venue for next meeting (whenever possible it is best to have this on the same day of the week at the
   same time, so that people will become familiar with the
   pattern)

Minutes: these are the official record of the business of the
meeting, related to the agenda, and are the responsibility of
the meeting's secretary. The secretary will have taken notes
of all discussion throughout the meeting, and these will be
typed and circulated to all concerned as soon as possible.

Refreshments: it is always a good policy to offer tea and biscuits
at meetings, for it is surprising how this brings out the timid
person during the more relaxed interval, with the result that
he or she will play a more productive part in the proceedings.

Venue: ensure that a suitable hut or room is arranged, with
sufficient tables and chairs for everyone.
    The room should be well ventilated, light and clean, to
create the right atmosphere.

Filing

An important facet of communication is the storage of
information. This covers a wide range of sources, from
computers and micro dots to the ordinary manilla folders.
    We often have to look up data, refer to letters and documents
of all kinds, and it is the essential of a good filing system that
the documents we require are kept safely and can be found
quickly. Generally, documents are filed in one of three ways:
1 Alphabetically

2 Numerically
3 Subject in a building organisation.

Alphabetical filing: a comparison can be made here with a telephone directory, for names are simply arranged in alphabetical order. Selecting the first letter to start with, and subsequent letters in the name if two or more are alike, e.g. Adams would be filed before Amos. This is an easy form of reference as a separate index system is unnecessary.

Numerical filing: with numerical filing, an index system is required. Its main advantage is that additional references can be added without having to rearrange the system, by just starting a new folder and adding a number to the card index.

Subject filing: similar to alphabetical filing but now information is classified by its subject, e.g. bricks, bonding agents. Although both begin with the letter 'b', these would be separated in a different file.

Whatever type of filing system is used, either at head office or on site, the same general principles will apply:
1 System should be simple to understand and use.
2 Information should be obtained quickly.
3 Should be compact to save space.
4 Should be economical to install and maintain.
5 Should allow for growth in relation to future needs.

Office appliances

Just as builders are becoming more mechanically minded regarding use of plant on site, so is the use of office machinery being made more and more. The type and amount of appliances to be found in a builder's office will largely depend upon the size of the organisation.

Calculating machines: very widely used and basically required to save time and mental effort in the process of addition, subtraction, multiplication and division and many other tasks. The information is recorded generally either in a small aperture on the machine or on a tally roll.

The degree of training required to use these machines varies with their complexity.

Typewriters: different makes of typewriter can produce a wide range of print. The three main types of machine used by the builder are:
1 Standard machines—will do the bulk of office work
2 Electric machines—powered by mains electricity, use approximately the same amount of power as a standard house lamp. These are faster than standard machines and are less tiring; the resulting work is generally of better appearance. Programmable machines are also available.
3 Portable machines: ideal for taking to meetings, light in weight, with carrying case.

Duplicating appliances: again, the type and amount of work required to be carried out will dictate the type of machine that is wanted from the many available.

Stencil duplicating: the print is typed on to a stencil, which is of wax-like composition. The stencil is placed on an inked drum; the ink will pass through the punched letters, leaving an imprint on the sheet of paper with which it comes into contact. Stencils can be cut by hand using a stylus (similar to a dry ball-point pen) with exactly the same results. Stencils can produce up to 5000 copies at one time, or can be stored carefully to be used again at a later date. Absorbent paper is used with these machines, and if the spread of ink is correctly controlled, the results are of a high standard.

Offset litho duplicator: the 'stencil' in this case is a thin metal sheet upon which information is recorded by hand or type-written, and then 'secured' by means of a chemical wash. The main sheet is wetted by rollers, and will not, therefore, pick up the oily ink, whilst the impressed letters reject water and take the ink, thus producing print. Produces up to 50 000 copies, but machine is expensive.

Spirit hectograph: a big advantage with this method is that colours can be used. Special carbon paper is placed under the master document, which is also a special paper. The impression is made by the use of a colourless liquid, which combines with the carbon to leave a print. Up to 200 copies can be taken from the master, and this results in an inexpensive reproduction.

Photo copiers: like computers, have changed considerably over the past years thanks to developments in the field of electronics.

Copies produced by the newer machines are of a very high quality and relatively inexpensive, the copy being produced in very few seconds. Many machines are now able to increase or decrease the size from that of the original.

The biggest advantage with this system is that production of a stencil or other form of master is not necessary. Reproduction can be made direct from most documents—the only restraint may be size, although many machines now in general use can cope with A6 to A3.

Photo copiers are often hired and if maintained in good order can work for long periods with little or no attention.

Purchasing new equipment

When considering the purchase of new office equipment, especially machines, many points should be considered before final selection is made to ensure that the item will be of economical value. Typical items that may require consideration include:

a What exactly will the new machine do?
b Which, if any, items will it replace?
c Overall cost of similar equipment (value for money).
d Will staff need to be retrained or new staff taken on?
e Is there room for machine? (check on alterations that may be necessary, consider floor loading, ventilation etc.).
f Period and cost of maintenance.
g Ease of repair (should be given special consideration if foreign equipment is to be purchased).

These items suggest areas of consideration but will of course vary considerably with type, size and purpose of equipment that is required.

19   Book-keeping and finance

There are many fine books on book-keeping and accountancy and it is not the purpose of this chapter to compete with them; the aim is to give a basic background of what accounting is all about so that the advantages and a simple understanding of them can be realised and the terms used understood.

Anyone, in any type of business, whether a sole trader, partnership, private or public company, will wish to know at some stage of time, generally annually, how they stand financially. By taking a critical look at the financial accounts, many items can be ascertained, possibly the most important being: has a profit or loss been made? Other points that can be ascertained are:

How much trade has been carried out over a set period.
What has been the cost of running the business—overheads.
What is the working capital of the business.
What discounts were paid or obtained.
How much plant and stock is there in hand,
and many, many more important and salient points that it is important to know in order to ensure that a business is running efficiently and in an economical manner.

Each and every business that keeps books of accounts may go about it in any way they choose, each firm selecting various books of accounts and systems of entering details to suit their own needs. The builder, for example, will not only wish to know how well he has done (annually) but will also wish to know how each contract or job fared so that he can ascertain profits or losses on any particular job. He would therefore be advised to keep independent contract accounts. In brief, therefore, it can be stated that no business enterprise can successfully operate without some system of accountancy, preferably under the direction of an accountant.

It is sad to reflect that many small traders go out of business each year, builders forming the biggest group, and often this is caused simply because their accounting system was at fault or in some cases non-existent.

Day-to-day business

Daily transactions are taking place in a business all the time and it is this basic problem that makes book-keeping necessary so that they can be controlled and records kept of them. An outline of the system of book-keeping, showing how these daily transactions are taken into the accounts and, over a period of time, result in the forming of the business's financial statement, is shown, Fig. 59.

Original entry

As can be seen from Fig. 59 the daily transactions pass through books termed books of 'original entry'. These books form the gateway to the ledgers and enable items to be collected and passed into the respective ledgers in lump sums and not as a huge list of small entries or transactions.

To illustrate this point an example can be found when dealing with petty cash. Most offices and sites require a certain sum of money to enable them to make ready cash payments for small (petty) items. If all these transactions had to pass through the cash book it would soon become an arduous task for the book-keeper to keep apace of them. To avoid this problem a set sum of ready cash is allocated to the office or site. This enables these transactions to be dealt with quickly and efficiently. All transactions must be recorded. The only sum concerning the book-keeper will be the set sum allocated, not the numerous small items.

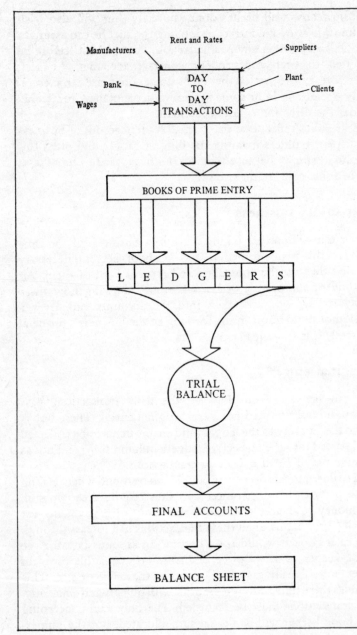

Fig. 59

Petty cash system

Just as in any system when cash is being handled it is necessary for records to be kept, petty cash is no exception. This is quite often the responsibility of a young clerk or trainee supervisor (technician) on a site and forms a good training for later work with money and costs. The method of keeping records relating to petty cash is known as the 'imprest system', the 'imprest' or 'float' being the allocated fixed amount. No money should ever be paid from petty cash without a receipt being obtained in exchange, consequently at any time the total value of receipts or vouchers, plus the cash balance (what's left), will all tally with the original advance.

PETTY CASH VOUCHER — PLUS — CASH BALANCE EQUALS IMPREST SUM

Most firms provide petty cash vouchers as illustrated in Fig. 60; alternatively 'chits' or 'receipts' must be presented.

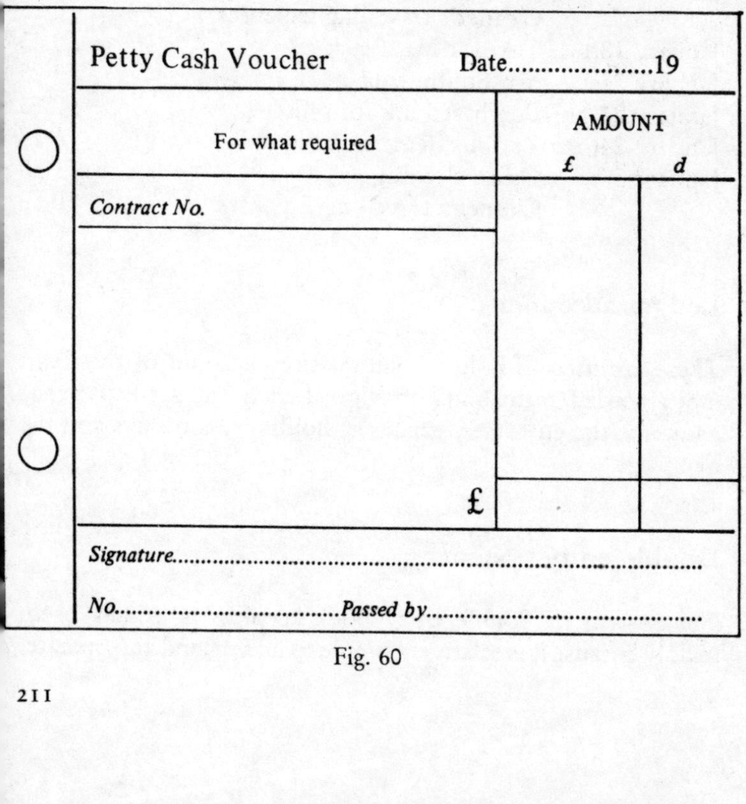
Petty Cash Voucher Date......................19

For what required | AMOUNT £ d

Contract No.

£

Signature...

No.....................................*Passed by*...

Fig. 60

Weekly or monthly the total expenditure is compiled and passed to the appropriate department, where it is checked and a covering cheque is issued, restoring the balance to the previously allocated sum.

The totals of the respective analysis columns will at this period be debited to the various accounts, the names of which head the columns. Example 5 illustrates a typical petty cash book.

January 1st	Received imprest cheque	£30·00
January 1st	Bought postage stamps value	£1·50
January 1st	Paper, envelopes, etc.	£1·00
January 1st	Paid by cash, cartage of parcel	£0·55
January 4th	Office cleaned	£1·50
	Cleaners travelling expenses	£0·15
January 6th	Writing paper	£0·27
January 9th	Postage stamps value	£2·00
January 12th	Parcel sent by post	£0·30
January 14th	Office cleaning	£1·50
	Cleaner's travelling expenses	£0·15
January 18th	Taxi for Mr. Black	£0·43
January 21st	Mr. Smith, haulage of goods	£3·25
January 21st	Purchased ink for office	£0·13
January 24th	Various items of stationery	£1·27
January 30th	Office cleaning	£1·50
	Cleaner's travelling expenses	£0·15

Ledger accounts

These are used to hold the summarised account of the items first recorded in the book of original entry and are kept in the names of the customer, expenses, holdings, contracts and the like.

Double entry system

The system of double entry book-keeping is widely used, mainly because it is relatively simple to understand and operate.

PETTY CASH ACCOUNT | ACCOUNTS TO BE DEBITED

Imprest £ p	Date 1979	Details	Vou. No.	Postage £ p	Stationery £ p	Haulage £ p	Travelling £ p	Cleaning £ p	£ p	TOTAL £ p
30 00	Jan 1	To cheque	—							
	Jan 1	By Postage stamps	1	1 50						1 50
	Jan 1	By stationery	41		1 00					1 00
	Jan 1	By Haulage	42			55				55
	Jan 4	By Off. cleaning	43					1 50		1 50
	Jan 6	By Writing Paper	44		27					27
	Jan 8	By Postage stamps	—	2 00						2 00
	Jan 12	By Parcel Post	45	30						30
	Jan 14	By Off. cleaning	46					1 50		1 65
	Jan 18	By Taxi	47				51			51
	Jan 21	By Smith Ltd	48			3 25				3 25
	Jan 21	By Ink	49		1 13					13
	Jan 26	By stationery	50		27					27
	Jan 30	By Off. cleaning	51				15	1 50		65
30 00	Jan 31	By Balance c/d	c/d	3 80	2 67	3 80	86	4 50		16 65
14 35										14 35
15 65										
30 00										30 00
	Feb 1	To Balance	b/d							
	Feb 1	To Cheque								

Example 5

The system works on the principle that if one person or one account gives, then another person or account must receive; this brings about a double 'entry' in the records because every transaction will be entered twice.

What now has to be decided is not only what ledger the amounts will be posted to, but also on which side, for each account will have a debit and credit side. The golden rule here is:

Credit (Cr)—the giver
Debit (Dr)—the receiver

Example: take the following day-to-day transaction

1	Paid rates by cheque	£150·50
2	Bought stock on credit from ABC Suppliers	£263·77
3	Paid cartage of goods to contract No. 1 by cash	£7·61

How are the double entries made?

1a	Rates account	Received	DR
b	Cash book	Gave	CR
2a	Purchase accounts	Received	DR
b	ABC Supplies	Gave	CR
3a	Contract No. 1 account	Received	DR
b	Cash book	Gave	CR

These items would be shown in the ledger. Example 6 shows how item 3a and b would be dealt with. The double entry has been made; the books balance. This is the essence of the Double Entry system.

At the end of certain periods, these accounts would be settled, whether the business owes or is owed money. This again must be a double entry. Taking the example already used, assuming that the client of Contract No. 1 paid by cheque for the delivery 3 days later, it would be entered as shown on p. 199.

It can be seen that a double entry has again been made, the books still balance, and entries can now start afresh.

Note: It must be remembered that it is more than likely that many items would appear in the Contracts account and that not each item would be paid for separately. On a large contract the balance would be made each month by the issue of an interim certificate on the monthly valuation, but the procedure is the same.

LEDGER ACCOUNT

CONTRACT No.1 ACCOUNT

Date	Details		£	p	Date	Details		£	p
Jan 13	To carriage of goods	CB	7	61					

CASH BOOK

Date	Details		£	p	£	p	£	p	Date	Details		£	p	£	p	£	p
									Jan 13	By cheque	CL					7	61

CONTRACT No.1 ACCOUNT

Date	Details		£	p	Date	Details		£	p
Jan 13	To carriage of goods	CB	7	61	Jan 16	By cheque for carriage	CB	7	61

CASH BOOK

Date	Details		£	p	£	p	£	p	Date	Details		£	p	£	p	£	p
Jan 16	To cheque	CL1					7	61	Jan 13	By cheque	CL1					7	61

Example 6

Books and accounts

The various books that may be used in the construction industry are as shown on pp. 201 and 202, together with their rulings.

Trial balance

At intervals, varying with the business but often annually, the books are closed. For example, if it was decided to close the books on 31st December as many of the outstanding accounts as possible should be settled prior to this date. The aim is to assess how the business has progressed during the trading period and how the business stands, considering all its assets and liabilities. Before the accounts are presented in their final form, a check is made on the accuracy of the ledgers by producing a trial balance. Using the double entry system of book-keeping we have seen how each transaction is made into a double entry, one entry on the credit side, the other entry on the debit side; it should therefore follow that the total entries on both the debit and credit sides should be equal—provided that no mistakes have been made. It is, of course, impossible to guarantee that no errors have been made and it is the purpose of the trial balance to check on the accuracy of the book-keeping.

TRIAL BALANCE
AS AT 31st DEC 1971

Date		FOLIO	DR £	p	CR £	p
Dec 31	Cash in hand	CB	150	00		
	Cash in bank	CB	3150	00		
	Freehold Premises A/c	LR	5000	00		
	Stock A/c	LR	430	00		
	Plant A/c	LR	2170	00		
	Capital A/c	LN			15250	00
	Insurance A/c	LN	150	00		
	Telephone A/c	LN	75	00		
	Salaries A/c	LN	4325	00		
	Discount A/c				200	00
			15450	00	15450	00

JOURNAL

Date		Fo.	DR £	p	CR £	p

As all entries must go through a book of prime entry a journal is necessary to record transactions which cannot be entered in either the Cash Book purchases—Sales or Return Books

PURCHASE DAY BOOK

Date	Name	Details	Folio Dr Cr	Total £ p	Stock £ p	Cont.1 £ p	Cont.4 £ p	£ p	£ p

On collection at the end of a trading period the ledgers (typical heading shown) are debited. The Person giving the goods is credited.

MATERIALS OUTWARDS BOOK

Date	Name	Folio Dr Cr	Total £ p	Cont.3 £ p	Cont.4 £ p			

This book records the stocks that have been despatched to contracts. The stock account at the end of the trading period will be credited with the total value sent out. The various contract ledgers will be debited for amount received.

WAGES ANALYSIS

Date	Folio Dr Cr	Total £ p	Office Salaries £ p	Yard Wages £ p	Cont.1 £ p	Cont.2 £ p			

Shows how the wages are dispersed The cash book will be credit with total amount. The various accounts being debited with their respective sums.

217

CASH BOOK

Date	Detail	F O L I O	Discount Col. £ p	Cash Col £ p	Cheque Col £ p	Date	Detail	F O L I O	Discount Col £ p	Cash Col £ p	Cheque Col £ p

Records all payments into and out of the banks —shows the ledger account cash transactions including all discounts received and allowed

Other books of prime entry may be used include:

PETTY CASH BOOK
PLANT ISSUE BOOK—Items will be Cr. in the Plant A/C   Dr. to contract were sent
PLANT RETURNED BOOK—Items Cr. to contract A/C   Dr. to plant A/C
STORES ISSUED BOOK—Items Cr. to stores A/C   Dr. to contract were sent
STORES RETURNED BOOK—Items Cr. to contract A/C from which stores returned   Dr. to stores A/C

ESTIMATES OUTWARDS BOOK

Date	Name of Architect to whom sent	Details of Estimate	Price	Completion Date	Result	Job No.	Remarks

ESTIMATES ACCEPTED BOOK

Date	Client	Address	Architect	Address	Price	Comp Time	Cont. Ledger No.	Remarks

The Estimate Books are kept to ensure that a close record is maintained of all possible and actual work and to ensure against overtrading.

LEDGER

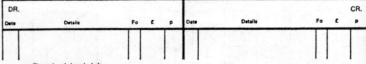

DR.									CR.
Date	Details	Fo	£	p	Date	Details	Fo	£	p

The various ledgers include

REAL ACCOUNT. Shows disposal of particular assets e.g. Stock. Furniture, Plant, Equipment etc.
NOMINAL ACCOUNT. Shows items used in running concern, Rates, Rents, Salaries etc.
BOUGHT ACCOUNT. Gives details from whom purchases have been made, Merchants, Suppliers etc.
SALES ACCOUNT. Shows to whom goods or work has been sold.
CONTRACT ACCOUNT. Necessary to indicate Profit or Loss.

The trial balance itself consists simply of collecting the balances of all the credit entries and comparing them with the balances of the debit entries. If the totals do not agree, it indicates that there is some error in the accounts which requires locating and correcting before the books will balance.

A typical example of a trial balance appears on p. 200.

All errors that may occur are not shown by the trial balance; these include:

1 Sums shown correctly as debits or credits but entered in the wrong ledger account
2 It will not show a total omission—when a transaction has been overlooked completely
3 The duplication of an entry—entered twice in error
4 Errors that compensate each other, errors having the same effect on debit or credit figures, but wrong

Errors it will show include:

1 Omission of one entry, debit or credit
2 Entry on the wrong side of a balance
3 Incorrect entries on either side, e.g. Dr £17·10 Cr £7·10
4 A misreading through poor figures

Final accounts

These are necessary to show all informative and useful information in respect of the monetary results of business over a particular period of time. Assuming that the ledger balances are correct as shown by the trial balance, the Final account can now be produced, all information being taken from the trial balance. The reason for taking out final accounts is basically to see if a profit or loss has been made. Any business has certain assets and liabilities; the business will over a period hope to increase its assets, not only possibly in cash, but, like a builder, in land, plant, equipment and the like. At the end of a trading period, the excess of the assets over the liabilities will be the capital.

Trading account

This account is used to ascertain the Gross Profits: that is, the difference between the purchases and the sales. e.g. if materials and work on a contract totalled £3500 and the contract price was £4000, then the gross profit of £500 has been made.

It must also be remembered that not all the purchases may have been sold; this will be shown in the stock account; the total amount shown on the closing date of the stock account will have to be credited to the trading account, whilst in the next trading period the value of this stock becomes the opening stock on the debit side, as illustrated in worked example which shows that £500 was left from the preceding trading account, and that stock to the value of £5700 is now left.

Stock valuation

Care must be taken when considering the true value of stock remaining; valuing too high or too low will result in an inaccurate gross profit in the trading account. It is more than possible that purchases will change in price over a period and to overcome this problem one of two methods of assessment should be used: use the cost price of the goods, or the current market price, whichever is the *lower*.

Profit and loss account

This account is again prepared from the trial balance plus the gross profit from the trading account. We have, through the trading account, ascertained the gross profit, which is the reward for effort in business. It is not, however, the true profit of the business. To discover what this is, all other expenses such as salaries, rates, rent, heating, lighting, insurance, etc., generally referred to as 'overheads', must be deducted; this will of course reduce the gross profit to the final or net profit earned. This is why it is called the profit and loss account.

By referring again to the worked example, it will be seen that all expenses and losses are debited, whilst all income and gross profit is credited.

If the credit side is greater, a net profit has been made; it the debit side is greater, a net loss shows, reducing the capital of the business.

Balance sheet

The balance sheet is not an account but a 'statement' of the business's assets and liabilities, which will show the exact financial position of the business at a stated date, indicating whether it is solvent or insolvent. If the total assets are greater than the liabilities—excluding the capital—the business is solvent, and vice versa. Capital cannot be taken into account as an asset because it is owed by the business. See worked example.

Value Added Tax

V.A.T. was introduced on 1 April 1973 with a standard rate and zero rate together with some exemptions. The Customs and Excise office is generally responsible for the local admini-stration of V.A.T.

Briefly, V.A.T. is charged to the taxable persons by his supplier on the goods and services they supply to him for his business, including capital goods, trading stock, materials for use in manufacture, and services used in the day to day running of the business. Those goods and services are called his 'inputs' and the tax on them is his 'input tax'. When he in turn supplies goods and services, not necessarily the same ones, to his customers, he charges the customers with V.A.T. The goods and services he supplies are called his 'outputs', and the tax he charges is his 'output tax'.

H.M. Customs and Excise will require returns at regular intervals when the input and output tax amounts are totalled and the smaller being subtracted from the larger, the difference is the amount he has to pay to, or H.M. Customs and Excise repay.

Zero-rating in construction
Generally speaking new work is zero-rated and applies to registered traders of services and materials in the course of construction.

Standard rated

Basically work that falls within the area of repair or maintenance is standard rate. Often in construction, new work and alterations and repairs are carried out together. In these circumstances if the alteration is a separate piece of work or can be separated from the rest of the job, it can be zero-rated. If it cannot be separated, e.g. alteration from repair or maintenance work, then the entire amount is liable to a positive rating.

Depreciation

Earlier in the chapter the accurate assessment of remaining stock was explained, the reason being that a true picture of the state of the business is necessary and wrong assessment of stock would show an unreal state in the business accounts. The same situation will also occur with other material items that may fluctuate in price over a period of time. A classic example of this is in the buying of a new car. If the purchase price was £10000, after one year one may own a £10000 car but its true value may only be £9000 owing to the fact that certain wear and tear will have taken place, so no one would now give the original selling price.

It is therefore reasonable to assume that if a business has £100000 of new plant and equipment at the beginning of 1987, its value by the end of the year will be considerably less and, like stock, a true realistic assessment must be made of the new value so that it can be accurately recorded in the accounts. If the original purchase price is shown it will show the business (assets) worth more than they really are.

Decrease in value

Loss in value is generally caused in one of three ways:
1 Wear and tear, as already indicated: this occurs due to the normal working life of a purchase.
2 Obsolescence: this occurs when purchases are superseded in time by more up-to-date items.
3 Wasting asset, such as a sand quarry, when the value is

bound to be reduced owing to reduction of material with working over a period of time.

Some of the items a builder will have to consider when deciding on depreciation values include:

1 Plant—mixers, cranes, lorries, diggers
2 Equipment—power hand tools, planners, circular saws
3 Office equipment—printing machines, adding machines
4 Office furniture—desks, chairs, filing cabinets
5 Property—store-rooms, offices—if leasehold, as value will decrease as lease runs out.

It can be seen from the above list that many items will lose value over a period of time due to wear and tear or obsolescence. If the business is going to continue, these will need to be replaced and so to provide for this replacement a builder will set aside a sum of money each year to accumulate over a period of time, enabling him to purchase new items without having to take lump sums out of profits or capital. The life span of items ranging from concrete mixers to filing cabinets is bound to vary. Therefore most builders vary the amount of money put aside each year so that too much capital is not tied up. The amount put aside is generally calculated from the life expectancy of the item, for whereas a calculating machine in an office may last 10 to 15 years, a hard-working concrete mixer would last approximately 5 years.

The methods adopted for accumulating the lump sum to purchase new items are as follows:

a A fixed percentage on diminishing value: this is a useful method of assessing depreciation for, as shown in the example, as the item gets older, the amount of depreciation gets less; this compensates for maintenance and repair costs which will undoubtedly rise.

Example: plant cost £1000—depreciation 10% per annum.

Original cost of plant	£1000
Less depreciation in 1st year 10%	100
Value of plant, end of 1st year	900
Less depreciation in 2nd year 10%	90
Value of plant end of 2nd year	810

etc., etc.

b Fixed or equal instalment method

This is calculated by determining the life of the item and simply dividing this into the original cost.

A scrap value may be taken off original value as shown.

Example: cost of plant £1000, estimated life 10 years: scrap value £100.

Original cost of plant	£1000
Less scrap value	100
	900

Therefore amount of depreciation deducted would be £900 ÷ 10 = £90 per annum.

c Re-valuation of assets

This method is usually adopted for such items as wheelbarrows, shovels, sledge-hammers and other such small items, generally classed as non-mechanical plant. Simply, items are listed and a new value is given to each; if the new assessment is lower than the original this would be written off in the profit and loss account.

Examples of how depreciation is shown in the final account can be seen in the worked example and how it is used in calculating plant hourly rate, Chapter 13.

Terms

Assets: Items that can be found on the assets side of the balance sheet; these indicate the gross worth of a business.

Current assets: the term used to illustrate the items that form the entries that are used in day-to-day transactions and fluctuate due to normal trading, e.g. stock, money in hand or in the bank, etc.

Fixed assets: these consist of items necessary to carry out the functions of business but are not bought or sold in the normal day-to-day transactions, e.g. premises, office equipment, land, plant, etc.

Intangible asset: goodwill comes under this heading as being worth something or that has a value but could not be sold like a tangible asset.

TRADING ACCOUNT OF A.N.OTHER

For Year Ending 31st Dec 1979

DR   CR

1970					1971				
Dec	31	Opening Stock	500	00	Dec	31	Sales - Daywork & Completed Contracts	30,000	00
Dec	31	Work in Progress	5000	00	Dec	31	Closing Stock	5,700	00
1971									
Dec	31	Purchases & sub-contractors	14800	00					
Dec	31	Wages and National Insurance	8000	00					
Dec	31	Power, light and heat. Joiners Shop	800	00					
Dec	31	Carriage and haulage	1050	00					
Dec	31	Gross Profit transferred to P & L A/c	5550	00					
			35700	00				35700	00

PROFIT AND LOSS ACCOUNT OF A.N.OTHER

For Year Ending 31st Dec 1979

Dec	31	Salaries	2250	00	Dec	31	Gross Profit	5550	00
Dec	31	Rent and Rates	150	00	Dec	31	Interest at Bank	60	00
Dec	31	Repairs to Premises	700	00	Dec	31	Discounts Received	240	00
Dec	31	Light and heat in Offices	290	00					
Dec	31	Reserve for bad and doubtful debts	100	00					
Dec	31	Postage and Telephone	120	00					
Dec	31	Depreciation of Plant and Equipment	200	00					
Dec	31	Net Profit	2040	00					
			5850	00				5850	00

BALANCE SHEET OF A.N.OTHER

As at 31st Dec 1979

LIABILITIES					ASSETS				
Sundry Trade Creditors					Cash at Bank			2000	00
D. White Ltd	2500	00			Sundry Trade debtors				
P. Green Ltd	2000	00			Dark windows Ltd	3500	00		
C. Bee	1200	00			Less res. for bad and doubtful debts	100	00		
			5700	00				3400	00
Capital	6000	00			Stock			5700	00
Add net profit	2040	00			Plant and Machinery	2840	00		
			8040	00	Less depreciation	200	00	2640	00
			13740	00				13740	00

<u>Order of Liabilities</u>

    These are shown in the order in which they should be paid

<u>Order of Assets</u>

    These should be arranged in order of easiest return of cash. This is called order of liquidity.

The date is important on the balance sheet since it will only indicate the standing of the business at this time, owing to the fact that due to trading the position of the business will constantly fluctuate.

Liabilities: indicate all outstanding debts for which the business is legally bound to pay.

Current liabilities: Items that the business must pay in the immediate future occurring due to day-to-day transactions, e.g. monthly trade credits, temporary bank loans.

Fixed liabilities: capital owing to the owner or partners or long term loans.

Types of businesses

Sole trader: business owned by sole proprietor is still commonplace, especially in the building industry. Anyone can start his own enterprise as long as he has a place from which to operate and capital to purchase goods so that he can sell or use them for further gain. The main thing is, of course, that any profit or losses that occur are his responsibility and his alone. If the firm fails, he, as the proprietor, is liable in full to his business creditors to his last penny—in other words, unlimited liability.

Partnership: when a sole trader wishes to expand he may decide that the best method of obtaining capital is to go into partnership. Not more than twenty persons can join together to form a partnership firm, and however many there are, an agreement should be drawn up legally and signed by each of the partners, setting out clearly the conditions that will apply within the firm to suit themselves. An example of a partnership agreement is shown:

Partnership agreement
1   The name of the company shall be: A. N. Other Construction, registered under the Registration of Business names Act unless actual surnames of partners are in name of firm.
2   The business shall be for the purpose of: Small Constructional Work to the value of £75 000
3   The address of the company shall be: Redhill, Warwickshire

4   The commencement of the partnership shall be 01 March 1988 for a duration in the first instance of ten years

5   Any retiring partner shall give 6 (six) months' notice

6   The capital shall be maintained at the following:

J. Kay	£3 000
T. White	£2 000
D. Green	£2 000
Total:	£7 000

Additional capital may be invested at 5% per annum but no interest be paid on initial capital

7   No interest to be chargeable on drawings or advances but in any one year drawings by any one partner must not exceed a net value of £2500

8   Partners' salaries shall be subject to annual revision, initially £720 monthly

9   Any profit or losses shall be shared in direct proportion to capital investment

10 The Trading period shall be 01 March to 28 February

11 Books of the company shall be audited annually by Frost and Cold, Accountants

12 Any partner found to be guilty of unprofessional or infamous conduct shall, at the discretion of the other partners, be liable to forfeiture of his rights in the partnership from any date specified.

13 The death of any partner shall not constitute dissolution of the partnership

14 New partners will only be introduced upon agreement of all existing partners

15 The dissolution of the partnership shall only be by unanimous agreement of all partners

16 Any insoluble dispute by the partners shall be decided by arbitration

17 Partners will act in the designated posts as listed:

J. Kay	Contracts Co-ordinator
T. White	Plant and Sales
D. Green	Estimator and Safety Officer

All will have equal right in the making of policy decisions.

If, however, the parties cannot agree as to what conditions should be in the agreement (Deed of Partnership) the law provides Statutory Control under the Partnership Acts of 1890, 1907, which provide that:

1 Partners share equally in profits and losses and capital.
2 Partners receive interest at 5% yearly on loans they make to the firm.
3 Receive no interest on their capital
4 No salaries to be paid

Each partner, as in the case of a sole trader, is liable to unlimited liability regarding the debts of the partnership firm, including his private possessions.

Types of Partners: Not all the partners in a partnership are classed under the same terms. These include:

1 Sleeping (or dormant) partners: these are partners who have contributed capital but wish to take no part in the daily running of the firm.
2 Active partners: these actually see to the running and general supervision of the affairs.

Partnership accounts

The method of book-keeping for a partnership is the same as already explained, the only difference being that the partner will have extra ledger accounts as follows:

1 Capital account
2 Interest on capital account
3 Drawings account
4 Interest on drawings account
5 Partners' salaries account (if salaries are paid)

The other difference between the sole trader and the partnership firm is in the final accounts. This is termed an appropriation account. This is an additional sub-section to the profit and loss account and will show on the credit side the net profit calculated and on the debit side the division between the various partners, as shown in the following diagram.

Date	Detail		£	p	£	p	Date	Detail		£	p	£	p
Dec 31	To Partners salaries	J					Dec 31	By Nett Profit				2500	00
	Black		500	00									
	White		500	00	1000	00							
	Interest on Capital												
	Black		100	00									
	White		75	00	175	00							
	Profit shared												
	Black ½ of £1325		662	50									
	White ½ of £1325		662	50	1325	00							
					2500	00						2500	00

The limited company

We have looked briefly at two sorts of ownership of businesses:
1 The sole trader
2 The partnership
Now to consider the larger companies. These are generally formed to accumulate sufficient capital to operate in a big way, requiring large capital resources to develop and expand.
These companies fall into two main classes:
1 Public company
2 Private company
Both are covered by statute law under the Company Acts of 1948.

The basic differences between the two are:
1 Any seven or more persons can form a Public Company whilst any two up to a maximum of 50 can start up a Private Company.
2 Whereas shares in a Public Company can be bought or sold freely, the transfer of shares in a Private Company is restricted.
3 The Private Company cannot offer shares to the general public.

229

The process of keeping accounts with Limited Companies is not very much different from the method already described only inasmuch as:

    a the method of raising and recording the capital, and

    b the form of the Final Accounts.

A Public Company is formed by registering it with the Registrar of Joint Stock Companies of London. A certificate of Incorporation will be issued by the registrar if the necessary documents required by law are satisfactory. These are:

    a The memorandum of Association, which must state:

      (i)   Name of Company (ending with Limited)

      (ii) the address of Registered Office

      (iii) object of the Company

      (iv) a declaration that liability of members is limited

      (v) the amount of Share Capital and the shares into which it is to be divided.

    b Articles of Association

This is the agreement which sets out the regulations between shareholders and the regulations that govern the general internal working of the Company: in brief, the rules. When the Company becomes limited, it means in effect that members will only be liable for the total amount of the shares they have taken out, e.g. if a shareholder subscribes for 100 £1 shares, then the extent of his liability is £100. Money is raised by the issue of shares up to the value of share capital shown in the Memorandum of Association.

Shares

There are various types of shares but generally they are classified as:

1   Ordinary Shares

2   Preference Shares

As the name suggests, a preference shareholder has a priority of any dividends relating to profits available but only up to a fixed % rate. The ordinary shareholder, however, is not restricted by this fixed dividend so that in good years when the profit is high, he will receive a good return, although in bad years he may get nothing.

Other types of shares are:

Cumulative Preference Shares: as the name implies, the dividend accumulates each year when profits are not sufficient to pay out; when profits become available, the total amount will be paid, i.e. cumulative profit.

Preferred Ordinary Shares: are shares that acquire their dividend after Preference shares, but before ordinary shares.

Deferred Shares: These will only receive dividend after all other shares have been paid.

Selling shares

The object here is to sell the shares for the highest price that can be obtained, either privately or on the Stock Exchange. If strong interest is shown in a company, then the shares will obtain good prices from prospective investors. A Register of shareholders is required to be kept so that changes of ownership can be recorded.

Debentures

Here is another method of obtaining money.

These are issued in addition to shares and are loans to the company, generally, but not always, repayable on a specified date. The big difference between a debenture holder and a shareholder is that a fixed interest is received regardless of whether a profit has been made or not.

Accounts

A profit and loss account together with a balance sheet must be presented by the directors of Private and Public Companies, the Balance Sheet containing a summary of share capital of the company.

The requirements of these accounts are laid down in the Companies Act, 1948. Companies must also hold an annual general meeting, when auditors must be appointed.

The bank

The banking systems used in this country can be classified
into 2 main groups:

1 *The Bank of England:*
   This is not a bank for the ordinary man in the street to
   use, but is designed to control the monetary situation of
   the land.
   Possibly the best known duty of the Bank of England is
   to issue the currency of the realm. It also controls through
   the Minimum Lending Rate the general interest paid on
   mortgages and loans. The Bank of England also keeps
   accounts, not of the normal saver, but of Government
   departments and other banks. It is obvious, therefore, that
   it holds a commanding role in the field of banking and
   national economy.

2 *Commercial Banks:*
   These are the banks that not only look after the individual
   but also transact much of the business done by commercial
   and industrial undertakings. The services these banks offer
   cover the complete field of cash transactions, a few of
   which are listed:
   a Provide a safe place not only for money but other valu-
   ables such as documents, jewels, etc.
   b Pay fixed sums when required, through the system of
   'Bankers Orders' to specified persons.
   c Will provide the correct amounts of various coinage so
   that wages can be made up.
   d Will, if sufficient security (and property, etc.) is available,
   provide loans and overdrafts.
   e The bank, generally the manager or a senior member of
   staff, will offer advice as to business ventures, shares,
   investments and the like.
For general day-to-day bank transactions, a choice of two
accounts will be offered:
   a Deposit Account
   This is when a sum of money is not required to be used
   for some time; it can be placed in the bank not only for safe

keeping but as the bank will use this money for their own purposes they will pay interest on the amount deposited. Money, except in small amounts, cannot be drawn from this account at any time and a cheque book will not be issued. A set period of notice is required for large cash withdrawals.

b Current Account

A cheque book is issued with this account as it allows money to be transferred at any time, allowing business to be transacted without the necessity of carrying around large amounts of cash, as a cheque issued is a request to the bank to pay the bearer the sum indicated. For this service no interest is given, in fact a charge is made for each cheque drawn.

Banks are limited companies, trading in money and security, and one of the banks' main form of income is from interest charged on money loaned. A loan is a fixed sum given to the applicant for a specified period of time; on this sum, interest is paid.

If, however, the depositor is temporarily short of funds, he may ask the bank for an overdraft. This is a short term measure, when the bank allows cheques to be drawn on an account they know has no funds available. Security will be necessary to obtain this service; interest rates are generally high and a maximum drawing figure is given.

20 Computers

The computer is, and will become even more so, a 'tool' of management and an aid to administration. Although it is not necessary to have a detailed knowledge of how a computer works or what it can do, it is nevertheless important to have a background knowledge and understanding so that the benefits and the advantages that they can offer will be fully utilised to aid the company.

History

The concept of computers is not a twentieth-century idea, for in the mid-seventeenth century, a young Frenchman by the name of Blaise Pascal developed a digital machine for aiding calculations. Then in the nineteenth century an Englishman called Charles Babbage developed these ideas and built a machine that automatically calculated mathematical tables; but it was not until engineering techniques were more advanced and world wars made it necessary that a practical start was made on Babbage's theories.

Dr. Hollerith, an American, developed to a very high degree the use of punched cards, first introduced in the eighteenth century by a Frenchman, Joseph Jacquard, who used punched cards in the silk-weaving industry. Dr. Hollerith used punched cards to help and speed up the national census. His electrically run punched card machine soon caught on in industry. The success of the punched card machine (electrical computer) delayed the progress of the electronic computer and it was not until the early twentieth century that the electronic machines started to develop. Since then, the growth has been steady and increasing in pace with the passage of time and the gaining of knowledge, often through the research carried out in universities, with the second world war producing big advancements, right up to the present time.

Computers have been classified as they have progressed, as follows:

1st generation—valves used
2nd generation—transistors used
3rd generation—micro integrated circuits

Developments mean that we now also have mini- and microcomputers.

Introduction

For building firms to survive and grow successfully they require two essential qualities: they must be both efficient and competitive. This applies not only to site working, but also to in-house administration.

In an industry renowned for generating vast quantities of information to be stored, retrieved and communicated, efficiency can be a nightmare.

The competitive edge is only gained when costs are lower than those of competitors when procuring work (including company overheads). Thus, if the work of office staff can be more speedy and efficient, then the company can be more cost effective and therefore more competitive.

As an administrative tool, the modern computer is second to none.

What is a computer?

Although it is unnecessary for users of computers to understand how they work, a brief analogy may clear away some of the mystery which surrounds them.

In simple terms, a computer can be likened to a massive bank of pigeon holes in which individual items of information can be stored and subsequently retrieved. The same information can be manipulated, either in word or number form, then re-stored as new information or presented to the user as results. This is no special achievement in itself, until the computer's greatest asset is introduced—its speed. Large computers can manipulate about one million items of information *per second*.

Allied to this, computers never get tired, do not require coffee breaks nor pay rises, only 'strike' when there is a power failure or when the programs which run them fail. Altogether, a model employee.

Who can afford them?

In the early days of computers when electrical technology was developing more slowly, only the very large building firms could justify the kind of expenditure necessary to buy computers.

The development of the cheap microprocessor has meant that computing power is available to *all* firms, from the very largest to the sole trader.

Microprocessors

The heart of the modern computer is inexpensive because the world has a wealth of the material from which it is made. Pure silicon is extracted from sand and each crystal is 'grown' to a rod of about 1 m length and 100 mm diameter. This is then ground smooth, cut into 'wafers' of about 1 mm thickness. Onto the wafer is printed the circuitry of the processors—often more complex than a street map of London—at the rate of about 50 circuit sets per wafer. The wafer is then cut down into 'chips'.

Compare the complexity of the electronics of one chip—the size of a small fingernail—to that of a complete valve-driven television of old and it becomes more easy to understand the phenomenal reductions in size and cost which have been achieved. Bear in mind also that the initial design of a chip may take ten man years or more, thus, volume sale is essential for economic production.

Computing equipment

With the 'jargon' of computers being the reason for the premature end to many conversations where 'non-computer' people are concerned, the briefest list of necessary terms follows:

1 *Hardware*: can be considered as any item requiring an electrical power source to make it work. Hardware includes:

   a *The computer*, i.e. the unit itself, whether 'mainframe' (very large), 'mini', or 'micro' (decreasing in size with respect to ability and storage capacity).
   b *The keyboard*: normally attached to the computer. Laid out in much the same way as a standard typewriter, it is the user's means of direct communication with the computer.
   c *The monitor*: a television with only one channel—tuned to the computer. It is one of the computer's outlets, with which it can communicate with the user.
   d *The disc drive*: a machine which reads magnetically stored information from 'hard' or 'soft' discs in much the same way as a tape recorder reads music from tapes—though much more quickly.
   e *The printer*: a computer driven typewriter printing about a line of text per second. Terms like 'dot-matrix' and 'daisy-wheel' indicate the manner in which the text is generated.
   f *The plotter*: much the same as the printer but normally associated with the production of drawings. Plotters producing drawings up to AO size are becoming fairly commonplace.
   g *The modem*: this stands for 'modulator/demodulator' and converts computers signals into signals which can be transmitted through the airwaves or through telephone lines, thus allowing remote computers to 'speak' to each other.

2 *Software*: essentially the catalyst which allows interaction between computer and user; commonly known as programs (yes, in computing, 'programs' *is* spelt correctly!).

Here, it must be noted that computers are quite stupid and

useless until they are prompted by commands from the user or from programs.

Also they do *exactly* as instructed provided that the instructions they are given are *totally* logical and in a language which is understandable to them. Thus, a program is a series of ordered commands which the computer can understand and to which it may respond, including the calling up of information from the user via the keyboard.

3   *ROM*: stands for 'read only memory' and is the set of fixed instructions built into a computer which never go away, even when the machine is switched off.
4   *RAM*: stands for 'random access memory' and is, effectively, the size of the pigeon hole bank available to the user and the programs used. The RAM capacity is one of the deciding factors a firm will take into account when purchasing a computer.
5   *Eprom*: stands for 'erasable—programmable read only memory'. This is usually an add-on component which may be integrated into smaller computers to allow specialised peripheral activities to be carried out. Typically, word processing is an eprom activity. Where typists normally type below their optimum speed to prevent or avoid errors and use carbons or a photocopy machine for multiple production, so the computer word-processor encourages the use of maximum speed and accepts error. The user may then browse through the letter or document on the monitor screen, edit it, re-arrange it, add or delete words, paragraphs or pages and then have as many copies as required produced in a range of type styles by the printer. The text can be saved on disc if of a repetitive nature. This example typifies the way in which eproms can benefit business.
6   *Networks*: to close the section on computers in general, mention has to be made of the current trend in computing. To provide *each* person in an office, or on site with hardware for computing would be very expensive and totally unnecessary. Instead, one master computer is used to feed information to, and be fed data from 'terminals'. A terminal is generally only a computer and monitor arrangement which has access to a large master machine. Information from the master can

be manipulated at the terminal and the results fed back to the master for further action or storage. The combination of the large master machine and the relatively inexpensive terminals linked to it permits many people or even separate departments or businesses to have massive computing power cheaply. A further benefit is that a piece of information generated at one terminal can be processed by the master and automatically transferred to 'interested' workstations. This must be the ultimate in communication, especially in construction.

Computer applications

The building industry has many procedures which are repetitive. Not only the laying of bricks and other physical activities, but also design and administrative processes. Any series of activities which can be resolved into a repetitive list is suitable for computing. Software can be specifically designed to allow the more mundane elements of the process—data storage and retrieval, data manipulation and mathematical calculations—to be carried out quickly and accurately.

The following is an overview of some of the applications for computers in construction. Every day, new ideas are being developed and no text can ever stay abreast of reality.

1 *Land surveying.* The power of the microprocessor has taken away the need for pencil, paper and calculator. The most modern instruments take and store all the surveyed information, just by the pressing of buttons. The time spent on site by the surveyors is dramatically reduced.

   If this is a benefit, then the real blessing shows in the surveyor's office. The survey instrument's processors, with the stored survey information, can be linked to a large plotting machine. This will very rapidly draw out the survey showing all the items the survey took in, things such as boundaries, buildings, roads, railways, trees—in fact *anything* that is required to be shown.

   The modern plotter presentation is of very high quality and the accuracy of the plot is extremely good.

2 *Design (architect)*. Part of the skill of the architect is to interpret the client's requirements. This interpretation is traditionally offered as 'sketch design'. As the client's ideas and the architect's impressions work towards agreement there may be a great deal of work for the officer eraser!

Some quite remarkable computer modelling packages have recently been developed, which allow architects to 'build' a realistic three dimensional image of a proposed structure on a monitor screen. This can be done very quickly, as can modifications. In fact, the structure can be 'tailored' to suit the client's requirements in his presence. The more sophisticated programs allow the building to be viewed from any angle, with respect to its surroundings and both from outside *and* inside. Furniture, furnishings and even people may be introduced onto the graphic image to show the client as near an approximation of reality as possible.

An example of the diversity of this kind of modelling occurred when one particular client, having given requirements for a house specifically requested that the house be positioned so as to obtain the most sunlight possible. The computer was set to work, rotated the house through all possible plan angles with respect to all possible sun positions for the whole of one year. The client was shown the most beneficial position of the house *the same day*!

A further benefit of computer aided design is, once again, on the plotting table. Where a few years ago design drawings generated by computer were considered 'stark', refinements are now allowing very acceptable working drawings to be produced. As design is built 'on the screen', the computer recognises dimensions with respect to the screen image and ensures that the total of internal compartments dimensions add up to the correct overall external dimension. It can also ensure that when an adjustment is made to a detail of design *all* things affected by the one adjustment are automatically adjusted themselves. This is a massive benefit in modern structures where the provision of services is a labyrinth of criss-crossing pipes, wire and ducts.

It is guaranteed that the development of computer aided design and its subsequent use will ultimately benefit building in general.

3   *Design(Structural Engineer)*. It has been said by some that architects draw pretty pictures and structural engineers make them work. Certainly, structural engineers are charged with the task of ensuring that the arrangements of beams, columns, floors, walls and roof provide the levels of stability and safety required by legislation.

These, therefore, are men and women with the knowledge of the mathematics and of the rules of strengths of materials. It needs no more than to reiterate two words: 'mathematics' and 'rules', to come to a rapid conclusion. We are back in the domain of the computer.

Straightforward design, for example, of timber roof trusses, portal frames, etc. can be analysed by computer extremely quickly. Complicated frames may take a little longer, but certainly not the time required by the human mind, even aided by an electronic calculator.

Perhaps the largest benefit is that, due to its speed, the computer can be set to consider alternative structural designs (steel, reinforced concrete, combinations of the two) in order to resolve not only *feasibility*, but, of equal importance, the most *cost-effective* system.

4   *Design (Services)*. This area must now be considered as possibly the most complex element of the design of a large building.

The sizing of pipes, cables and ducts is straightforward in computing terms as, like the structural engineer, the services engineer also works with mathematics and sets of regulations.

More importantly, computer modelling allows the engineer to 'inspect' the layout on the monitor to ensure that each pipe and conduit has a unique and not a shared position.

Computers can bring forward the day when we will no longer see a plumber and an electrician racing for the right to an opening large enough for just one service!

Preparation of Bills of Quantities

The task of accurately taking off quantities for complex structures can be quite a daunting task. Nevertheless it is essential, for without Bills of Quantities, accurate tender pricing by contractors is difficult.

Each work item in a Bill consists of three elements:

a a description
b a unit of measure (m, m^2, m^3)
c the quantity of that work to be carried out.

Bill preparation can be carried out much more quickly when the computer is employed.

Quite simply, standard descriptions of work items are stored in a databank. These may be accessed and fed, with dimensions, into a process which resolves them into net quantities of the work to be carried out. The savings in both time and 'paper' can be staggering.

The contractor

Having opened the chapter with two of the contractor's needs for survival and growth—efficiency and the ability to be competitive, it seems prudent to close with the same kind of firm.

The range of administrative duties the computer is able to carry out is widening virtually daily and thus can only be briefly mentioned here.

Items such as planning (including CPM), cost control and budgetting, payroll, accounts, materials scheduling, estimating and many more elements fall within the widening range of computer abilities.

Applications with respect to the contractor are curtailed to a mere list, as to totally identify the way in which computing can aid a firm in its administration *could* occupy a companion volume of the same thickness as this book!

Conclusion

The final years of the twentieth century will no doubt see more and more advances in the range of abilities of the computer. Manager and technician alike must learn to accept and respect the machine for its awesome power yet still retain the faith that it cannot match the human mind for inventiveness—merely take the notions of great minds and implement them.

21   Training

The need for training concerns all, from management to operatives, and with the industry becoming more and more technical, competitive and complex, the need for training becomes more necessary. The old maxim of combining theory with practice still holds good, and the only variance to this is the time span, for it can easily be understood that it will take a much longer period to learn and apply management techniques than to learn a practical skill.

The construction industry is broadly divided into three main groups:

1   Manual workers (tradesmen and labourers)
2   Technicians (the assistants to the professional men)
3   Management

Manual workers

Most of the present-day tradesmen served an apprenticeship varying from three to five years. The apprentice is registered by the National Joint Apprenticeship and Industrial Training Commission through BEC or BATJIC FMB, who now have their own schemes and will receive from them a registration card. The deed of apprenticeship is executed by the employer, the apprentice, his parent or guardian, and a representative of the

Joint Apprenticeship Committee for the district not earlier than the boy's sixteenth birthday. The system of training the apprentice by setting him to work alongside a skilled craftsman is still carried on, but is becoming more difficult with the growth of self employed craftsmen, and sub-contract groups.

The pattern of training has altered considerably over the past few years. Although many apprentices still attend College on the well tried method of one day and one evening per week, the industry's preferred method of training is through the CITB Foundation Training Scheme.

In this pattern of training a young person is taken on in the first year but is employed as a trainee, generally doing 24 weeks 'off the job' training at a college and working the rest of the time on site or workshop. At the end of this training year most are taken on as apprentices and follow a second year of training of between 7–12 weeks 'off the job', the time differing with the actual trade involved. At the end of this period the apprentice will take a City and Guilds examination and if successful will be awarded a Craft Certificate in this trade.

The craft certificate allows the student to do a further year's study, either day or block release, for an Advance Craft Certificate. Both certificates are awarded to show the proficiency of the apprentice in both theory and practice.

Many young tradesmen go onto further study in the field of technician education, become very well qualified and leave the 'tools' to take up posts of a supervisory nature.

With the backing of industry and through the CITB, usually about six months after taking the Advanced Craft the apprentice will have to take a Skill Test, designed to illustrate their capability in relation to site practice. These tests are all carefully designed and given rigid time limits to put candidates into an industrial situation. It is only after passing this test that the young person acquires the status of Craftsman.

Training of semi-skilled operatives and labourers has, under the Training Board, also been made possible, often being carried out at colleges or the National Training Centre at Bircham Newton.

Technicians

The BTEC (Business & Technician Education Council) courses which are the equivalent to the old Ordinary National Certificate and Diplomas are now firmly established.

These courses are designed for the young technician and lay the foundation for Higher Certificate and Diplomas, and in some instances Degree work. Like the City and Guilds courses they indicate standards of achievement, enabling employers to select the right person to fill a vacancy.

The design of this course allows for differing programmes of study to suit the needs and requirements of all technicians, giving a choice of subject matter through the various modules which go to make up a specific course. The nodular course also allows students to proceed at varying speeds, dependant upon their abilities, needs and standards of achievement.

Since September 1987 an integrated approach to the BTEC courses is available, enabling students to see, through projects, or similar work, how each modular subject is only a part of the whole framework of construction.

Technicians, like members of other sections in the industry, can progress to senior posts, management and professional positions.

Various institutions have been created to serve the technician, including the Society of Architectural and Associated Technicians (SAAT) and the Society of Surveying Technicians (SST).

Management

All persons who have supervisory duties are referred to as management, whether it be a director, top management, contracts manager, quantity surveyor or buyer, middle management or site agent, general foreman, craft foreman, site management. All have the same role to play, that of ensuring that the tasks they control are completed to the satisfaction and benefit of the company.

Generally, training for management is a long term exercise. In the building industry most management staff come from one of the professions, e.g. surveying, estimating, planning, etc., and through internal training and education courses learn to understand and adopt principles of management after having acquired a professional qualification. Many of the technicians have the opportunity to become middle management due to the tasks they perform and the standard of attainment required to pass their appropriate examinations.

Construction Industry Training Board

Established under the Industrial Training Act of 1964, its purpose is to ensure that there is an adequate supply of properly trained men and women at all levels of industry, at the same time improving the quality and efficiency of industrial training. To carry out these objectives, a levy was imposed, making all employers contribute and so share the cost of training. This sum is now to be levied under occupational categories, i.e. differing rates for the various occupations.

To recover money paid to the C.I.T.B., the employer may send apprentices, or other employees of any age, to attend approved courses at a technical college or other training centres, or may even set up his own training school for which he will receive grants varying with the standard of the course.

The Employment and Training Act 1973 modified the original Industrial Training Act and provided for the establishment of a Manpower Services Commission, an Employment Service Agency and a Training Service Agency.

22 Introduction to management

This chapter is designed to help the young technician and the aspiring manager to understand more clearly the relationship between administration and management.

The two are very closely related, one not being able to succeed without the other, or in fact there being no need for one without the other.

Management is the task of the manager, or leader. It is his role to formulate policies and ideas he/she feels are best for the organisation, whatever level of management this is – Prime Minister, Board of Directors down. When the manager has resolved his ideas he requires them to be put into practice, so he passes them on to his subordinates who must try to ensure that they are carried out in a satisfactory and financially profitable manner. This is done by the use of administration in all its forms.

In a preceding chapter we considered the attributes of a leader. We should now consider his role, firstly stressing the practising theories which are exactly the same whatever his standing or level of management, the only difference being the degree in depth of its application. For example, although a contracts manager's work is vastly different to that of a craft foreman, both are managers, both use the same principles. A craft foreman may have had little or no training in management and may, therefore, not know or understand the principles which he uses in everyday practice.

247

Management is not new and has been practised throughout history, whenever a leader has emerged or been appointed. In more recent times man has thought more about the art of practising management. This has, therefore, improved and made easier the work the leader has to do. This study of management gathered pace with the event of factory work and production, especially during the Industrial Revolution when problems in organisation brought an awareness of the situation, not least of the social considerations. No one person can be given the title of the founder of management for as mentioned it has been an on-going practice. However, certain names are worth mentioning, these are persons who have made a very outstanding contribution to management, and the student manager should have a realisation as to who they are and what they did. The people mentioned are in no way an exhaustive list.

Pioneers of management

Robert Owen (1771–1858): Very much a pioneer of personnel management and was one of the first to consider the human factor as being important to industry. The list of changes he made within his own surroundings were impressive and included reducing the working day to $10\frac{1}{2}$ hours, not employing children under 10 years of age, and providing meals in the factory. Not only in the factory area did he improve conditions, but also the workers' domestic conditions by providing houses, workers' low price shops and generally improving the area to make it attractive.

Owen's work towards general social reform was a contributing factor in the passing of the first Factories Act in 1819.

Mary Parker Follett (1860–1939): Made an outstanding contribution to management as a political and social philosopher. Much of her work was concerned with social relationships within groups, examining leadership and the problems of authority. She was a firm advocate that leaders are not only born but can be made through training in the understanding of human behaviour, which gives rise to her principle of organisation relying on co-ordination.

George Elton Mayo (1880–1949): Was concerned with human and social factors in industrial relationships, particularly those arising out of social groups. He noted that training was carried out for new technical skills required as changes took place, but no thought was given to the training of new social skills required to balance the changes in method of living, a condition similar to that of the 1980s with the introduction of the 'chip' and the accompanying unemployed.

Benjamin Seebohm Rowntree (1871–1954): Was very much concerned with the human approach to management and philosophy which he applied to his own family business, improving industrial welfare, not only at the Rowntree Chocolate Factory where he was labour director, but also during the First World War, when at the Ministry of Munitions he founded a new Welfare Department to assist employers doing war-work to resolve unfamiliar human problems. His outlook on employees was of considerable concern, resulting in his developing a medical department, day continuation school, works councils, family allowances, unemployment pay and many such contributions that are taken for granted today. Rowntree was also concerned with the relationship between managers and workers, and to encourage this relationship he set up a full-time, company-paid shop steward, a works council and a joint appeals committee for disciplinary matters.

Frederick Winslow Taylor (1856–1915): Considered to be the father of scientific management. (See page 147.)

Henry Lawrence Gantt (1861–1919): Was concerned with the human aspect of the labour force, but his main claim to fame was the graphic production chart known as the Gantt Chart. (See page 148.)

Frank and Lillian Gilbreth (1868–1924): Developed motion study. (See page 148.)

Henri Fayol (1841–1925): Possibly Europe's most distinguished

249

figure in relation to management, one of his foremost theories being that of administration as a separate function. The processes that Fayol considered vital for higher management to adopt he put forward in 1908 in a paper entitled 'General Principles of Administration'. These principles being forecasting, planning, organising, commanding, co-ordinating and controlling, principles that today lay down the modern approach to problems of management.

As already mentioned, Fayol did much to develop the management principles and it is largely due to him that the process of management evolved – so much so that whenever consideration is made to the theory of management, the activities that he considered vital are included in the seven major processes of management.

The seven major processes of management

Forecasting: This is normally the area in which the Board of Directors or head of the firm are involved and it is their task to look ahead and forecast future developments so that the organisation may maintain a competitive business. Consideration must be given to working capital, the organisation structure, changes in the firm, production format, etc. Failure to forecast correctly could mean disaster for the firm, or at best, little development with only marginal profits.

The planner preparing master programmes must also forecast any difficulties or problems that may arise during the project period. Even the craft foreman must forecast by looking ahead (for example, to see what work his team will carry out in the next two or three days).

Planning: When a forecast has been made it is essential that the ideas are interpreted into a plan of action that can be worked to. The manager preparing the conditions and strategy that will enable the forecast to be fulfilled in all its many aspects, therefore, must give considerable attention to detail. The craft foreman will plan his workforce so that he can have people available to carry out the tasks that he foresees will need to be done.

Organising: Before any action can start all the resources, such as manpower, materials, plant, etc., must be marshalled to enable the ideas at planning stage to become a reality and carried out. This is a difficult area and requires excellent communication routes, relying very much on the administrative sources to see that nothing is forgotten that may cause delay, or upset, the planned system of work.

The craft foreman has the same problem, making sure everything is available for his team to carry out the work he planned they would be doing. He must ensure that all materials, scaffold, plant, etc., are available.

Motivation: A section dealt with under the chapter on organisation. A difficult area for it involves people and the art of getting them to want to work. Often referred to as commanding, directing or instructing. Every leader has their own method of encouraging their team to achieve good productivity, by knowing the conditions that help to motivate the individual members. The craft foreman will know his team as individuals and will most likely, without knowing it, achieve sound results by applying his personality to get them to work. He may also be able to relate the work to a sound incentive scheme which is a ready-made form of motivation.

Co-ordinating: This is a vital aspect of management and particularly in construction, as most work is carried out by small teams or gangs, which, if allowed to work without due thought to others, will result in delays and general chaos.

Careful balancing of the resources available is important at this stage, as nothing is achieved if one section of the work is completed in advance of other sections, without affecting the final completion date. If the craft foreman is in charge of brickwork he will have to consider the needs of carpenters, especially at first fixing, electricians, and plumbers and heating engineers for service ducts, etc., keeping to schedule so that plasterers and other finishing trades are not kept waiting. All this can only be done by co-ordinating each section, a problem generally resolved for him at the weekly site meeting.

Controlling: This does not mean the control and discipline of members of the workforce, but the comparison that has to be made between what was planned and what has been achieved (or the estimated performance against the actual performance). In the situation of planning or estimating, checking is done to see how one has achieved the standard laid down in time, output and costs, when the task has actually been completed. The craft foreman can easily see what work has been completed in relation to the work planned. If in any area when actual work is checked against estimated work and found wanting, corrective measures may have to be taken to put things right. This must be done immediately so that the overall plan can be maintained. It is also important that standards achieved are recorded in some way so that they can help to up-date and improve the estimated standards the next time (feedback).

Communicating: None of the processes of management should be considered as watertight or closed compartments; in fact, if they were, nothing would happen as information would not be passed on to enable others to do their work. It is therefore essential to have a sound, accurate and well-defined communication system; it is imperative in all phases of any work zone for the transmission of information if the planned results are to be achieved. The craft foreman will have been given oral or documented instructions as to his task for a period of time, say a week, and he will have to communicate back such details as man hours, usually through time sheets, allocation sheets, etc., progress achieved, outputs, and many other similar items.

Many of these process words have been used in various chapters of this book and it is hoped that they now have a more precise meaning, together with the realisation that they can be clearly used to give a logical pattern of thought and action to any leader/manager, whatever his status. Management's main aim is in directing human activities, for nothing can happen without a co-operative workforce. This can be summed up in a very simple management definition as: ensuring that the right people do the right thing at the right time and in the right place.

23   Building institutes

The Royal Institute of British Architects R.I.B.A.
Founded in 1834 and given its first Royal Charter in 1837. The use of the word Royal was granted in 1866. Its main aim is to draw up and administer the Institute code of professional conduct and set examinations for the maintenance of architectural standards.

The institute is also involved by representation on many committees from Government level down, relating to building topics.

The Royal Institution of Chartered Surveyors R.I.C.S.
Founded 1868 and incorporated by Royal Charter in 1881. The Institution covers many surveying specialisation under its examination structure. Has two classes of membership: Fellow and Associate.

The Construction Surveyors Institute C.S.I.
Founded 1952 and was formed to cater for the needs of the builders surveyor with grades of membership being obtained through practice and/or examinations mainly in the fields of quantity surveying, financial analysis and procedures.

The Chartered Institute of Building C.I.O.B.
Founded in 1884 but its roots lay in the 'Builder Society'

formed in 1834. It is now the major qualifying association for building, its 'membership' examination being very management orientated being reached by a series of grades: Student, Technician, Licentiate, Member (M.C.I.O.B.), Fellow (F.C.I.O.B.).

Incorporated Association of Architects and Surveyors I.A.A.S.
Founded in 1925 to encourage and facilitate co-operation between architecture and surveying and has corporate membership grades of Associate and Fellow which are obtained through examination and experience requirements.

Society of Architectural and Associated Technicians S.A.A.T.
Founded in 1965, the society promotes the recognition and standing of technicians. The society is affiliated to the R.I.B.A.

The Society of Surveying Technicians S.S.T.
Founded in May 1970 to establish a recognised organisation for surveying technicians. The society sets out to promote nationally accepted standards of technical ability. Membership is possible to the R.I.C.S. by a bridging structure.

The building crafts also have their own bodies such as the Guild of Bricklayers, the Institute of Carpenters and the Registered Plumbers. Their main aim is to bring interested parties together to maintain and improve craftsmanship. This they do by holding meetings and lectures and by sitting on many advisory panels, for example City and Guilds of London Institute examination moderating boards.

These and the many other institutes regardless of their field of activity have the general objective of securing the advancement of their members by laying down standards of integrity and competence.

24   Terms

Arbitration

When a dispute arises and cannot be settled fairly and amicably between parties, the problem may be taken to arbitration. This is, in effect, a technical court, having the full backing of the courts behind it.

A third person, the arbitrator (often an independent architect or quantity surveyor), will hear both sides of the dispute, including any witnesses. He will give a ruling which must be adhered to, as in a court of law.

British Standards

Prepared by the British Standards Institution which provides appropriate standards for materials used in the industry, laying down acceptable tolerance. The Institute also provides Codes of Practice; these do, however, deal more with workmanship than with materials.

Completion certificate

Within a 3 month maximum period of all defective work

being made good, the architect will, if satisfied, issue a final certificate, thus allowing the contractor to obtain all outstanding monies from the employer.

Daywork

Work is often difficult to measure or value, e.g. repair work, demolition, work below ground. In these cases, the JCT Form of Contract states that this work can be priced through daywork. This is submitted on daywork sheets, an example of such a sheet is illustrated, Fig. 61. Labour, materials, plant and any other charges, including a percentage for overheads, would be required so that accurate values can be obtained. If rates for daywork are not included in the Contract Bill, the national schedules of daywork charges can be used. Daywork sheets must always be signed by architect or clerk of works.

Defects liability period

This is a period of time usually 6 months, unless otherwise stated in the contract, when defects, shrinkages or other faults, due to workmanship or materials, generally shown in an architect's schedule of defects, must be completed. A completion certificate will not be issued until they are finished.

Employer's liability insurance

The contractor takes out this insurance which indemnifies the employer so that in the case of any person being injured or killed, he will not be liable, unless the accident is due to neglect on his part.

Fluctuations

During the course of a project operatives' wages and mater-

A.N. OTHER LTD
1762 BLOCK ROAD
REDHILL

DAYWORK SHEET   No 1739

CONTRACT..WEEK ENDING..................19......

DESCRIPTION OF WORK

No.	NAME	Trade	1	2	3	4	5	6	7	Total Hours	Rate	FOR OFFICE USE			
												WAGES	EXES	M.A.	T.M.
									TOTALS					

MATERIALS

Date	Supplier	Quantity	DESCRIPTION	Rate	Office Use

PLANT		For Office		TRANSPORT		Office		SUMMARY		
Qty.	Description	Rate			hrs	rate		LABOUR		
								EXPENSES		
								INSURANCE		
								H.W.P.		
								M.A.		
								T.M.		
								MATERIALS		
								PLANT		
								TRANSPORT		

AGENT   ....   ARCHITECT   ....
FOREMAN   CLERK OF WORKS

Fig. 61

ials may increase, National Insurance may also increase or decrease. To ensure that the contractor is not made liable for these increases and also to ensure that the employer obtains any decrease due to him, the fluctuation clause in the R.I.B.A. contract requires prompt notice of any changes made to be brought to the attention of the architect. Accurate records must be kept of any increases or decreases so that adequate additions or omissions may be made to contract price.

Goodwill

This is the estimated worth of future earnings of a business. To obtain a true value is a complicated process as past earnings, future trends, etc., must be considered. When a business is being sold or a person is buying into a partnership, goodwill is taken into account when deciding the final settlement.

Handing over

Upon practical completion of a project, a meeting is generally held between the employer, architect and builder to:
1   Explain any special use of equipment to the employer, e.g. heating units, etc., at the same time handing over all technical data.
2   Any special maintenance information must be given by the builder to the employer
3   The owner will now take over insurances.

Interim certificate

This enables the builder to receive regular payments (generally monthly) so that he will not have to carry the heavy financial outlay of a complete project. The quantity surveyor prepares a valuation of:
1   Work properly executed
2   Total value of materials and goods (basically those required for immediate use, provided they are adequately protected),

generally with the help and co-operation of the builder. Retention is then calculated and deducted from the valuation.

The quantity surveyor will pass on to the architect the final valuation, upon which, if the architect is in agreement, he will issue an interim certificate, which is in effect an obligation to pay the builder its value.

Measurement of work done

This is necessary for the payment of bonus. A true and accurate record, taken by a competent person, of work done to date. Great care must be taken to ensure work is not measured twice.

National Building Agency (NBA)

A Government grant-aided, non-profit-making advisory body set up to improve techniques and help increase productivity.

Prime cost and provisional sums

These are approximate sums that the builder will add to his own tender estimate to cover work and goods supplied by nominated sub-contractors and suppliers. Prime costs cover work by nominated sub-contractors, including statutory undertakings. Provisional sums are used when cost cannot be fully detailed.

Retention

This is a sum of money deducted from each valuation up to a certain total limit (termed the Limit of Retention Fund). Both the percentage retained each month and the total limit of retention should be inserted in the contract. Generally 10% is the maximum retained each month, with 5% of the contract being the limit held. The object of withholding some of the contractor's money is to safeguard the employer in case things go wrong. The contractor will not get all the retention back until the project is completed satisfactorily. It is therefore to the contractor's advantage to get the project completed as quickly as possible. Half the retention is generally paid with issue of the Certificate of Practical Completion, the final payment being made upon the issuing of the Certificate of Completion of Making Good Defects or at the end of the defects liability period, whichever is the later.

Standard method of measurement

This lays down the principles to which work must be measured in the bill of quantities to ensure that they conform to the requirement of the JCT and the GC/Wks of Contract.

Time extension

This is time allowed or given to the contractor over the agreed completion date of the project due to delays arising from circumstances over which the architect, employer or contractor(s) had no control. One of the most frequent of these is exceptionally bad weather, the onus being on the word 'exceptionally'. Extension of time may be granted if the employer or architect cause delays, e.g. big variations from original project. This extension must always be given in writing.

Variation order

The employer or architect may require, during the course of a project, to alter items such as:
1   An error or omission of an item in a bill
2   the alteration of work, goods or materials
3   alterations due to legislation, i.e. building regulations and

so on; these variations will be measured by the quantity surveyor, prices being based on similar prices in bills or daywork.

The architect cannot just issue a variation to suit himself and the builder can object if he considers the variation out of context with the original contract, e.g. contract to build a house and a variation is given to change it into a block of flats.

Wages

The current weekly standard basic rates of wages for craftsmen and for labourers is fixed by the National Joint Council for the Building Industry in accordance with Rule 7 of the Council's Rules and Regulations together with provision for a guaranteed minimum bonus payment to be made.

Wages (guaranteed weekly)

Operatives are now guaranteed payment at the standard weekly rate of wages for the full normal working hours of each complete payweek of the period of employment, subject to certain conditions laid down in National Working Rules.

Written instructions

If an architect issues instructions, they should only be acted upon when in writing, otherwise the contractor may have difficulty in recovering any costs involved through carrying out instructions issued orally. A set procedure is set down in both the JCT and the GC/Wks Contract document, so that confirmation in writing can be obtained.

Index